T0300981

"Australia's energy transition is uniquely fascinating: cheap renewables are taking over a coal-dominated system, and the prospect of declining fossil fuel exports is set against the allure of future clean energy exports. These changes create complex dynamics at the regional level. Written by academics who engage at the 'coalface' and practitioners, this collection presents a unique set of insights into how energy transition can be achieved at the regional level."

Prof. Frank Jotzo, *Crawford School of Public Policy, Australian National University; Head of Energy, ANU Institute for Climate, Energy and Disaster Solutions*

"From 'impossible to possible' is a testament to hope and tenacity. Despite the climate wars that have held Australia back, this book shows the progress possible with the power of community and place based development. The lessons learned from these regions demonstrate that the support of and co-creation with workers and community along with Government support make the difference. Valuable reading."

Sharan Burrow, *former General Secretary of the International Trade Union Confederation*

"As a non-academic operating at the interface of research and policy I must admit that I like this book a lot. It is deeply anchored in evidence, yet relatively easy to read. It talks about general trends but also dives deeply into relevant case studies. And more importantly, it does not talk about prehistoric cases, rather focusing on the past 15 years – a period most readers can actually still remember!

As an advocate of fossil fuel transition based in Europe I am also very pleased to have a thorough piece on Australia. Understanding of Australia's transition in Europe (and also in the Americas and Africa) is largely anecdotal and patchy – so a freshly written comprehensive piece is very much needed.

We are now at a critical stage of the global energy transition. Many jurisdictions have analysed and engaged on various options for that transition. But we still lack a global exchange of well-documented regional cases where one can follow the entire journey, understand the drivers and impacts and see both the national and regional contexts. *Regional Energy Transitions in Australia: From Impossible to Possible* does precisely that for Australia and I hope it will be followed by studies on other jurisdictions."

Andrzej Błachowicz, *CEO, Climate Strategies*

"People decarbonise society, not technology. This book puts people into the picture, charting region-level experiences across Australia. It gives us a grounded understanding of what works, what doesn't, and why. It makes an appeal for participatory planning against top-down imposition, and puts widespread public support for renewables at the centre of energy policy. The book is a vital antidote to current technocratic and neo-liberal approaches that exclude the public and make transition vulnerable to the fossil fuel lobby. That antidote is urgent and necessary, and not just in Australia, as governments across the globe face a backlash against corporate renewables and big 'green' capital."

Prof. James Goodman, *Professor of Political Sociology and Director of the Climate Justice Research Centre, University of Technology Sydney*

"As the IPCC has again recently confirmed, limiting global warming to close to 1.5 degrees requires a rapid shift away from unabated coal consumption. This book provides a valuable contribution to this increasingly urgent task by exploring learning from Australia (a major coal producer and exporter) about the development and implementation of equitable policies for accelerating the phase out of coal."

Prof. Jan Minx, *Head of Working Group on Applied Sustainability Science, Mercator Research Centre on Global Commons and Climate Change, Berlin*

"This book offers a timely perspective on early Australian experiences in the transition away from fossil fuels, an imperative in the global response to climate change. The book is an insightful collection examining regional transition processes and responses which highlights emerging lessons that can inform responses by government and other actors as they embark on the complex process of successful transition both in Australia and elsewhere.

A unique selling point is the combination of academic and real-world analysis by those involved in the transitions on the ground, making a valuable contribution to the just transition and economic development literature and also offering practical insights for other coal regions where regional transformation remains a pressing issue. Case studies link the local specifics of place with emerging themes across different regions, and the political and socio-economic dynamics will resonate with those involved in coal transitions across the world. A must-read for those interested in seeing how different social actors, political economy, and socio-economics intertwine to shape transition in coal-dependent regions."

Jesse Burton, *Senior Researcher, Energy Systems Research Group, Department of Chemical Engineering, University of Cape Town*

"Australia is a key actor in the global climate debate. As a fossil fuel producer, exporter and a country highly vulnerable to the effects of climate change, what happens in Australia has global implications. This new volume sheds invaluable light on energy transitions unfolding at pace across five key regions of the country, showing both how and why change is possible. Accessibly written, empirically rich and policy relevant, this accomplished set of authors has done a great service in exploring how rapid and just transitions to renewable energy are possible even in the face of fierce resistance and powerful vested interests. It deserves to be widely read by policymakers, activists and academics alike."

Prof. Peter Newell, *Sussex University, author of Power Shift*

"This concise and compelling book tells the stories of energy transitions underway in politically challenging contexts across regional Australia. With its long history of contested climate politics, Australia has always provided a key and critical lens through which to understand the dynamics of social and economic struggles to move beyond fossil fuels. Exploring the drivers and constraints, opportunities and challenges faced as diverse regions and communities attempt to foster just transitions, this book provides a thought-provoking analysis for all of those interested in how we can make just energy transitions a reality."

Prof. Harriet Bulkeley, *Department of Geography, Durham University and Copernicus Institute of Sustainable Development, Utrecht University*

"*Regional Energy Transitions in Australia: From Impossible to Possible* is superb! The case studies included in this book provide critical insights and highlight pragmatic lessons learned for a just energy and economic transition that are relevant not only to Australian communities dealing with these changes, but to communities around the globe."

Heidi Binko, *CEO & Co-Founder, The Just Transition Fund*

Regional Energy Transitions in Australia

This book provides an accessible and critical appraisal of Australia's regional energy transition initiatives.

The book begins by situating Australian energy transition in the context of Australian and international debates about climate change and energy transition. It then explores how energy transition planning was made possible in Australia's regional energy heartlands even while public transition planning was impossible. The authors consider five case studies of key early transition initiatives in the Latrobe Valley (Victoria), Hunter Valley (NSW), Central Queensland (Queensland), Port Augusta (South Australia) and Collie (Western Australia). They explore how transition came onto the agenda and outline the key actors, decision points and actions. The authors critically assess the successes and failures of each initiative, drawing out key learnings for other regions. The book concludes by evaluating the key cross-cutting themes emerging from the five case studies and draws out the lessons they teach about how to achieve a just transition.

This concise book will be of great interest to students and scholars of energy transitions, climate action, social justice, economic renewal and regional transition challenges and strategies, both in Australia and overseas.

Gareth A. S. Edwards is Visiting Associate Professor in the School of Global Development at the University of East Anglia, UK and Visiting Fellow at the Sydney Environment Institute, University of Sydney. He was the recent holder of a Leverhulme International Fellowship for research on justifications for ongoing coal extraction in Australia and India. He also led a British Academy-funded project 'A just transition away from coal in Australia' which sought to understand what 'just transition' means in Australia, the challenges

Australia will have to overcome to achieve a just transition away from coal, and the opportunities for reframing just transition ideas in ways which stimulate productive discussions between different stakeholders and communities.

John Wiseman is Senior Research Fellow at Melbourne Climate Futures and Adjunct Professor at the Melbourne School of Population and Global Health, University of Melbourne and Chair of the Board of The Next Economy. He is the author of numerous articles, book chapters and reports on climate change policy and energy transitions. His most recent book is *Hope and Courage in the Climate Crisis* (Palgrave Macmillan, 2021).

Amanda Cahill is CEO and Founder of The Next Economy, a non-profit organisation that supports regional communities across Australia to build more resilient, climate safe and socially just economies. Amanda has supported all levels of government, industry, workers and community groups to manage the energy transition across Queensland, the Hunter Valley, the Latrobe Valley and the Northern Territory. Amanda sits on the Australian Energy Market Operator's Social Licence Advisory Council and the National Hydrogen Strategy Advisory Council, and is a Senior Research Fellow at Melbourne Climate Futures at the University of Melbourne, an Industry Fellow at the Sydney Policy Lab and a 2023 Churchill Fellow.

Routledge Studies in Energy Transitions

Considerable interest exists today in energy transitions. Whether one looks at diverse efforts to decarbonize, or strategies to improve the access levels, security and innovation in energy systems, one finds that change in energy systems is a prime priority.

Routledge Studies in Energy Transitions aims to advance the thinking which underlies these efforts. The series connects distinct lines of inquiry from planning and policy, engineering and the natural sciences, history of technology, STS, and management. In doing so, it provides primary references that function like a set of international, technical meetings. Single and co-authored monographs are welcome, as well as edited volumes relating to themes, like resilience and system risk.

Cross-Border Renewable Energy Transitions
Lessons from Europe's Upper Rhine Region
Edited by Philippe Hamman

Energy Transition in the Baltic Sea Region
Understanding Stakeholder Engagement and Community Acceptance
Edited by Farid Karimi and Michael Rodi

Organizing the Dutch Energy Transition
Edited by Hans van Kranenburg and Sjors Witjes

Regional Energy Transitions in Australia
From Impossible to Possible
Edited by Gareth A.S. Edwards, John Wiseman and Amanda Cahill

For more information about this series, please visit: www.routledge.com/Routledge-Studies-in-Energy-Transitions/book-series/RSENT

Regional Energy Transitions in Australia

From Impossible to Possible

Edited by
Gareth A. S. Edwards,
John Wiseman and Amanda Cahill

First published 2025
by Routledge
4 Park Square, Milton Park, Abingdon, Oxon OX14 4RN

and by Routledge
605 Third Avenue, New York, NY 10158

Routledge is an imprint of the Taylor & Francis Group, an informa business

British Library Cataloguing-in-Publication Data
A catalogue record for this book is available from the British Library

ISBN: 978-1-032-85486-1 (hbk)
ISBN: 978-1-032-85509-7 (pbk)
ISBN: 978-1-003-58534-3 (ebk)

DOI: 10.4324/9781003585343

Typeset in Times New Roman
by Newgen Publishing UK

Contents

Acknowledgements

Gareth Edwards acknowledges the support of a Leverhulme Trust International Fellowship (IF-2021-021) and a British Academy grant under its 'Just Transitions to Decarbonisation in the Asia-Pacific' Programme (JTAP210019).

John Wiseman acknowledges the support of The Next Economy and Melbourne Climate Futures at the University of Melbourne.

Amanda Cahill acknowledges the support of The Next Economy.

The editors and authors acknowledge that this book was written on unceded Aboriginal lands and pay our respects to Elders past and present. No transition is just without the participation of Australia's first peoples.

Contributors

Stevie Anderson lives in Collie on Wilman Boodja.

Georgia Beardman is Research Assistant at the Centre for People, Place and Planet at Edith Cowan University on Wardandi Boodja.

Linda Connor is Emeritus Professor of Anthropology at the University of Sydney. She has conducted long-term ethnographic research on coal mining, climate change and renewable energy in Hunter Valley communities and in South Australia. Her publications include *Climate Change and Anthropos: Planet, People and Places* (Routledge Earthscan 2016) and *Environmental Change and the World's Futures: Ecologies, Ontologies, Mythologies*, edited with Jonathan Marshall (Routledge, 2016). She is a co-author of *Beyond the Coal Rush: A Turning Point for Global Energy and Climate Policy?* (Cambridge University Press, 2020) and *Decarbonising Electricity: The Promise of Renewable Energy Regions* (Cambridge University Press, in press).

Kimberley Crofts is a researcher and service designer with over 25 years of experience in Australia, Asia and the UK. Her PhD (submitted 2024) investigated community roles in knowledge co-production for sustainability transitions in the Hunter Valley. Kimberley was formerly a principal at Meld Studios, one of Australia's leading service design agencies. She holds a Master of Information Design, a Graduate Certificate in City Planning, an IAP2 Certificate in Engagement and Public Training, and a Bachelor of Visual Communication.

Jaime Yallup Farrant is the convenor of the Climate Justice Union WA based on Whadjuk Boodjar.

Elianor Gerrard is a social researcher and community development practitioner. She has a PhD in Environmental Studies from the University of Tasmania, where her thesis explored the community experience of (un)just transitions in Port Augusta, South Australia and the Latrobe Valley, Victoria. Elianor currently works as a Senior Research Consultant in the Energy Futures team at the Institute for Sustainable Futures, University of Technology Sydney. Her research focuses on community participation, ownership and social justice in the energy transition. Prior to becoming an energy researcher, Elianor worked in education, communications and community development across Australia and in Spain and Indonesia.

Naomi Joy Godden is a Vice-Chancellor's Research Fellow at the Centre for People, Place and Planet and the School of Arts and Humanities at Edith Cowan University on Wardandi Boodja in southwest Western Australia. She has a PhD in Social Work from Monash University. Naomi engages in feminist participatory action research (FPAR) with social movements in Australia, Asia and the Pacific to collectively understand the intersecting injustices of environmental change and develop and implement actions to demand feminist responses in policy and practice. Naomi has 20 years of experience in community development and ecofeminist research and activism with grassroots community organisations, local government, international development agencies, universities and the United Nations. She currently facilitates a range of FPAR projects in areas such as social justice and energy transitions, girl-led climate activism in the Pacific, Indigenous women's human rights movements in Australia, and integrating climate justice in community services.

Warrick Jordan is a born and bred Hunter Valley local. He has worked for 20 years on regional economic and environment issues, including providing career support for resource sector workers, establishing education and training initiatives, participating in successful multi-stakeholder environmental initiatives, and undertaking economic and social research. Past roles include as the Hunter Region Employment Facilitator for the Australian Government, the Coordinator of the Hunter Jobs Alliance, a unique union-environment alliance focused on practical responses to the energy transition and as a consultant economic geographer at the University of Newcastle.

James Khan is a traditional owner of the Wilman tribe known as the freshwater people.

Lisa Lumsden has 14 years of personal, academic and community practitioner experience in energy transition. She served as Port Augusta City Councillor between 2010 and 2018 during the surprise closure of the coal power stations and the subsequent rise of renewable energy. Between 2012 and 2017, she was a key volunteer leader in the 'Repower Port Augusta' community campaign calling for a fair and just energy transition and for her efforts is a recipient of the Jill Hudson Award for Environmental Protection. She has worked with the University of Technology Sydney on a transnational research project and is a co-author of the resulting book: *Decarbonising Electricity: The Promise of Renewable Energy Regions* (Cambridge University Press, in press). She currently works at The Next Economy and was instrumental in delivering the groundbreaking *Gladstone 10 Year Economic Roadmap* (2022) and the *What Next? Community Perspectives on Latrobe Valley's Energy Transition* (2023) projects.

Angus Morrison-Saunders is Professor in the Centre for People, Place and Planet and the School of Science at Edith Cowan University on Whadjuk Boodja.

Keira Mulholland is a mother of five, including a child with additional needs who lives and works in Collie on Wilman Boodja.

Dan Musil is a PhD candidate at Western Sydney University. Using a range of action research methodologies, Dan's research explores possibilities for low-carbon transition and transformation, with a focus on worker-ownership, economic democracy and the Latrobe Valley, where he lives. Dan has been engaged in work on just transitions for over a decade, including as Secretary of the Earthworker Cooperative—an initiative to build more just, democratic and sustainable economies in Australia and beyond.

Mehran Nejati is Senior Lecturer at the Centre for People, Place and Planet and the School of Business and Law at Edith Cowan University on Whadjuk Boodjar.

Joe Northover is a traditional owner of the Wilman tribe known as the freshwater people.

Jayla Parkin is a single mother who lives in Collie on Wilman Boodja with her two-year-old daughter.

Liam Phelan is Senior Lecturer in the School of Environmental and Life Sciences at the University of Newcastle, Australia. Liam's research interests centre on two areas. The first is environmental studies and science, with a focus on climate risk governance, just transitions and complexity; and the second is tertiary education, with a focus on science education policy and practice. Liam convenes the Bachelor of Science programme at the University of Newcastle.

Leonie Scoffern is a community liaison and member of Council for the Shire of Collie on Wilman Boodja.

Evonne Scott is a Social Work Honours graduate from Edith Cowan University on Wardandi Boodja.

Phillip Ugle lives in Collie on Wilman Boodja.

Lynette Winmar is a traditional owner of the Wilman tribe known as the freshwater people.

1 Creating just regional energy transitions

Key challenges and debates in Australia

Gareth A. S. Edwards and John Wiseman

Introduction

Just a few years ago in 2021, the possibility of a just and rapid energy transition in Australia looked decidedly remote. Coal was still king, gas was on the ascendency and 'transition' was an unutterable word in public, particularly in fossil fuel-producing regions. In June 2021, an energy company executive who mustered the courage to mention the need to manage transition at a community forum in Central Queensland resigned amidst political turmoil. Australia's Liberal National Party (LNP) coalition government continued to firmly reject the case for an urgent transition from fossil fuels to renewable energy, despite the fact that when emissions from both domestic and exported fossil fuel use are taken together, Australia is responsible for approximately 5% of total global greenhouse gas (GHG) emissions while having only about 0.3% of the global population (Christoff, 2022).

At COP 26 in Glasgow in November 2021, the Climate Action Network was quick to award Australia its 'Fossil of the Day' prize after the Australian government delegation (which shared its official pavilion with the gas giant Santos) arrived with no new targets for 2030.[1] The best the Australian Government could do was to pledge A$740 million from the Clean Energy Finance Corporation (its 'green bank') to investigate Carbon Capture and Storage, a technology with very few operating examples and widely considered uneconomic for thermal coal power (Groesbeck and Pearce, 2018).

DOI: 10.4324/9781003585343-1

But behind closed doors, tentative discussions about transition were emerging amongst the labour movement, environmental movement and investor community, each of whom were increasingly convinced that big changes in energy production were coming and Australia could no longer ignore them. Transition, its implications, and its management were also increasingly being raised in private in some of Australia's key fossil fuel-producing regions as communities began to grapple with what a changing world (and changing climate) would mean for their way of life.

The first clear bells of change had sounded five years before Australia's dismal performance at the Glasgow COP, when on 3 November 2016 the French multinational Engie announced it planned to close its Hazelwood coal-fired power station in the heart of Victoria's Latrobe Valley. Hazelwood was the largest and oldest coal-fired power station in Victoria. It burned low-grade brown coal (lignite) which was fed directly to its boilers from adjacent pits. Hazelwood and the three other power stations in the Valley supplied the bulk of Victoria's electricity generation capacity. The plant was shuttered in March 2017, just 5 months after Engie's announcement. The sudden closure came as a shock to both the local community and the state government, sparking both to action.[2]

While Latrobe Valley developments attracted the most public attention, other Australian fossil fuel-producing communities also began grappling with what a transition away from fossil fuel energy production would mean for them during the decade from 2011 to 2021. Transition initiatives—some more formal and organised than others—emerged from discussions and collaborations between labour, community groups, environmental groups, industry and local and state governments (Cahill, 2022).

These regional energy transition initiatives were thrust into the spotlight following the Federal election of May 2022. According to the ABC's *Vote Compass*, climate change was the number one issue mentioned by voters in the election campaign, which saw the Labor Party led by Anthony Albanese elected to government alongside a significantly expanded group of Green Party (+3 seats) and 'Teal' (+7 seats) members of parliament (MPs). The 'Teal' candidates were independents who shared a similar platform which combined an economically conservative political agenda with strong advocacy for accelerating action on climate change and political integrity and accountability. They were supported by the 'Climate 200' political fundraising group and many of those elected unseated high-profile Liberal Party (conservative) MPs.

One of the first acts of the new Australian Parliament was to legislate a *Climate Change Act* which committed the Commonwealth Government to a stronger emission reduction target for 2030 and net zero emissions by 2050. The *Act* also integrated climate action into the criteria for some government spending. In the two years since the election, closed-door discussions about 'transition' have been replaced by increasingly bold whole-of-economy transition planning institutional structures. In May 2023 the Commonwealth Government announced the establishment of a National Net Zero Economy Authority tasked with guiding transition planning for the country.[3]

In this changed political climate, policymakers at all levels have joined communities, investors and industry in actively exploring models and examples of how to transition the energy system away from fossil fuels while providing economic benefits for regional communities. The debate about energy transition has suddenly become very public and prominent. But the pace of change has afforded little opportunity for reflection on the lessons that can be learned from regions which had already begun to confront the challenges and opportunities of a just and rapid energy transition before the political climate changed. That is where this book steps in.

This book

This book's objective is to begin to fill a gap in our knowledge of how energy transitions in Australia's regions *have already been pursued.* It does so by documenting and exploring how transition was put on the agenda in five regions during the decade before the 2022 election. The regions in focus are arranged roughly chronologically according to when their encounter with energy transition began, though it quickly becomes apparent in each of them that the transition away from fossil fuels is only the latest in a long history of 'transitions' the communities have experienced. Chapter 2 examines Port Augusta (South Australia), Chapter 3 the Latrobe Valley (Victoria), Chapter 4 Collie (Western Australia), Chapter 5 the Hunter Valley (NSW) and Chapter 6 the Gladstone region in Central Queensland.

The authors of each chapter include contributors who were personally involved in the transition discussions of the region in focus.

Each author team was given six questions as prompts to guide their appraisal of how transition emerged in their region:

1. Where and when did the transition process begin?
2. Who were the key actors?
3. What were the key decision points and actions?
4. What proved easy to achieve and what was hard?
5. Which key opportunities were missed and with what effect?
6. What are the key lessons from this regional transition experience?

These questions aimed to generate analyses which highlight the successes of the initiatives, reflect critically on their failures and pause to consider the serendipitous connections which facilitated them. While energy transitions are still underway in all of the regions and judging their 'success' or 'failure' is premature, taking stock now provides an important window to understand what has worked and what hasn't in regional transition initiatives in Australia. Taken together, these regional examples demonstrate that steps towards a just transition can be taken in regional communities even under the most challenging of circumstances.

In the rest of this chapter, we set the scene by arguing that climate change necessitates an energy transition and outline some of the key drivers of transition internationally, which we argue have also been reflected in Australia. We then consider why Australia was so late to develop a public conversation around transition. Australia's role as a significant fossil fuel producer and the long period of political and policy paralysis that accompanied the 'climate wars' of the 2000s and 2010s are the two key factors. We briefly introduce these interlinked economic and political considerations, which highlight the key role of justice in enabling the energy transition. We explore the imperative for a 'just transition' that is a key driver of our project, before introducing the regional initiatives to be examined in Chapters 2–6.

Drivers of the energy transition

It almost goes without saying that current energy transitions are driven by the urgent need to avert further human-induced climate change. Fossil fuel energy generates more than 75% of GHG emissions and there is now widespread agreement that a rapid phase out of fossil fuels is essential to keep global warming as close as possible to 1.5°C

(IPCC, 2023). Even achieving the less ambitious target of limiting warming to 2°C is estimated to require a third of known 2010 oil reserves, half of gas reserves and over 80% of coal reserves to remain unused. This would mean that 95% of Australia's coal would need to remain unburned (McGlade and Ekins, 2015).

The barriers to rapid energy transitions are similarly well known. They include negative impacts on national and regional economic growth and employment, loss of export income and taxation revenue and the political influence and economic power of fossil fuel producers, investors and workers (Jakob and Steckel, 2022). In recent years, however, support for an energy transition has grown internationally, driven by several key factors each of which also apply in Australia.

The changing economics of energy

Continued and rapid falls in the cost of renewable energy have been a major driver of support for an energy transition as they are fundamentally reshaping the economics of energy production and consumption. Between 2010 and 2021, the levelised cost of solar photovoltaic electricity declined by 88%, the price of onshore wind energy fell by 68% and the price of offshore wind fell by 60%. With advances in energy storage, renewable energy is increasingly the cheapest form of energy around the world (IEA, 2024a, 2024b). This includes in Australia, where renewable generation (including associated storage and transmission) is significantly cheaper than any fossil fuel generation, and where nuclear generation remains the most expensive source (CSIRO, 2024).

In 2023, the International Energy Agency (IEA) predicted that the share of global electricity produced by renewable energy will rise from 29% in 2022 to 42% in 2028 even despite increased coal- and gas-based generation driven by the war in Ukraine (IEA, 2024a, 2024b). Australia provided a case study of why renewable generation also improves energy security when the Australian Energy Market Operator had to step in to control runaway energy prices in the National Electricity Market which were driven by a combination of high international coal prices, a shortage of gas due to producers favouring exports and outages at a number of coal-fired power stations. While rising energy prices continue to create significant concerns for Australian households and industries, the relative price of renewables compared to fossil fuels is a strong rationale for energy transition.

Changing public sentiment driving increasingly ambitious government action

Public sentiment has also played a large role in accelerating the pace of change, with concern about the implications and impacts of climate change rapidly growing around the world. In 2021, 72% of citizens in advanced economies were 'somewhat or very concerned that global climate change will harm them personally at some point in their lifetime' and 37% were 'very concerned', up from 31% in 2015 (Pew Research Centre, 2021). Over 75% of respondents to a 2022 OECD survey thought that 'climate change is an important problem and that government should take strong action to reduce climate risks and impacts'.

In Australia, the *Climate of the Nation* report (Australia Institute, 2022) found similar levels of concern about the effects of climate change. Seventy-five per cent of Australians were 'concerned' about climate change and 42% were 'very concerned', up from 64% and 24% in 2017. The report also found that 79% of Australians believed that Australia's coal-fired power stations should be phased out, including 31% who thought they should be phased out as soon as possible. It is likely that the 'Black Summer' bushfires of 2019-20 were instrumental in shifting Australian views.

This growth in public concern both internationally and in Australia has been reflected in commitments to reduce emissions from various sectors, including education, finance and publicly listed companies. An Australian example is how the Business Council of Australia's views on climate change shifted dramatically between 2019 and 2022. In 2019, it described a 45% emission reduction target as 'economy wrecking', but by 2022, it saw a more ambitious 46–50% target as 'pragmatic and ambitious'.

Growth in public concern about climate change has also been reflected in increases to both the ambition and strength of government commitments to addressing climate change. By 2024, over 100 countries representing more than 80% of the world economy had announced targets for achieving carbon neutrality by around 2050 (Climate Action Tracker, 2024). The interventions to support these decarbonisation targets have included pricing mechanisms to increase the cost of carbon pollution, incentives and sanctions to reduce demand for energy and increase energy efficiency, incentives to expand renewable energy generation, reductions in subsidies for fossil fuel producers and regulation to limit emissions.

A realisation that energy transition can drive job creation

While currently implemented policies are sufficient only to limit warming to around 2.7°C by 2100 (Climate Action Tracker, 2023), support for the energy transition has also been driven by a growing realisation that it can drive job creation. The International Renewable Energy Agency (IRENA) reported that a million renewable energy jobs were added in 2022 alone to a total of 13.7 million, and it expected the transition to create an additional 40 million jobs in the energy sector by 2050 (IRENA, 2023a, 2023b). The IEA (2022) estimates that more than 30 million jobs could be created in clean energy, efficiency and low-emission technologies by 2030, while in the United States, investment flowing from President Biden's Inflation Reduction Act will create over 550,000 new jobs in industries producing renewable energy.

Australian climate action and energy transition advocacy groups were first to argue that the energy transition had notable job creation benefits, but business groups have increasingly followed. The 2021 Beyond Zero Emissions (BZE) report *Export Powerhouse: Australia's $333 billion export opportunity* argued that there is huge potential for low-cost renewable energy to drive investment in high value, energy-intensive export industries including green hydrogen, green steel and green aluminium (BZE, 2021). The BZE report also notes, 'to capture this growing momentum towards zero-emissions markets, Australia needs a cohesive industry strategy and an ambitious climate target to keep up with our key trading partners' (BZE, 2021, p. 5).

BZE found an unlikely ally in the Business Council of Australia, which argued in the same year that

> the momentum for moving towards net zero by 2050 is unstoppable. The pace and scale of change is accelerating globally. Australia is at a crossroads: we can either embrace decarbonisation and seize a competitive advantage in developing new technologies and export industries; or be left behind and pay the price.
>
> (BCA quoted in SBS, 2021)

Most recently, the Federal Government has picked up on these calls, announcing a $22.7 million 'Future Made in Australia' package in the 2024-25 budget. The transition to net zero emissions through shifting to renewable energy and associated growth in green energy exports is the centrepiece of this package.[4]

Drivers of Australian inaction

The previous section shows that most of the international shifts in favour of energy transition have been reflected in Australia, so it is reasonable to ask why Australia was so slow to embrace the opportunities of transition. The two primary reasons are economic and political, and they are closely interrelated.

Australia as a fossil fuel superpower

Australia is a major global fossil fuel superpower. Its recoverable coal reserves are the third largest in the world, at 75,428 million tonnes (Mt) of black coal and 73,865 Mt of brown coal (Geoscience Australia, 2024). While Australia is the fifth largest producer of coal behind China, India, Russia and the United States, its per capita GHG emissions from coal were higher than any other developed country in 2021 (Readfearn, 2022).

Approximately 20% of Australian coal production is for domestic use. Over 90% of domestic coal is used to produce electricity with the remainder used for steel production. While the contribution of renewables to Australia's energy supply continues to grow, coal still accounted for 59% of Australian electricity production in 2021, down from over 80% in 2000 (Clean Energy Council, 2022). In recent years, the inescapable reality has been that Australia's domestic power generation industry is moving away from coal, and it is doing so primarily on the grounds of cost rather than due to climate considerations, as ageing coal plants reach the end of their technical lives and are too expensive to maintain or repair. It was cost that led Engie to mothball its Hazelwood power station in 2017 (see earlier this chapter, also Chapter 3). Cost is the primary factor behind the fact that no coal-fired power station has been opened in Australia since Bluewaters 2 in Western Australia in 2010. Cost has also been decisive in the closure of 12 of the 34 coal-fired power stations in operation in 2010 by the end of 2017 (Burke et al., 2019), the closure of another 6 by 2021 and the slated closure of 7 of the remaining 16 by 2035 (Carbon Brief, 2022). The Australian Energy Market Operator (AEMO) expects 60% of the eastern seaboard's fleet of coal-fired power stations to exit the electricity grid by 2030 with the last remaining plant shut by 2038 (AEMO, 2024).

But Australia's domestic use of coal pales into insignificance compared with its role as a coal exporter. Australia is the world's largest exporter of metallurgical coal and the second largest exporter of thermal coal after Indonesia. Australian export coal mainly goes to Japan, China, India, South Korea and Taiwan (Edwards et al., 2022). While most coal markets are in long-term structural decline, China and India are more than offsetting this with growth in demand (IEA, 2023). This means that discussions about Australia's economy often come back to coal. Coal has contributed significantly to Australia's economy and balance of trade, particularly in the context of high coal prices in recent years.

In 2021, Australia also became the world's largest exporter of Liquefied Natural Gas (LNG), exporting 81.2 Mt (Geoscience Australia, 2022). The value of Australia's LNG exports increased from $30.5 billion in 2020-21 to $70.2 billion in 2021-22, largely due to price rises linked to the impact of Russia's invasion of Ukraine (Ogge, 2022). The economic significance of Australia's gas exports is largely behind the more recent narrative that there should be further development of gas as a 'transition fuel', a bridge between coal-dominant and renewable-dominant energy regimes in Australia. The 2023 'Net Zero Australia' Report argues, for example, that even with the most rapid expansion of solar and wind energy, gas will be required to meet energy supply gaps during periods of peak demand. But many climate and energy analysts argue that while gas may be needed as a residual fuel to manage peak demand gaps in the very short term, there is no case for any new gas investment except as for intermittent backup generation (Denis-Ryan and Morrison, 2023; AEMO, 2024).

Given its significance to the national economy, it is not surprising that climate change and energy politics have been so highly politicised in recent decades. The country is dramatically exposed to changing energy and climate policies of the key trading partners who import its coal and gas. If Australia's major trading partners implemented policies consistent with keeping global warming below 1.5°C, Australian coal exports would fall by 80% (Kemp et al., 2021).

This is compounded by the fact that while the coal and gas industries do not employ large numbers of people relative to Australia's labour force, workers are overwhelmingly concentrated in a small number of regions where the industry dominates employment. Take the example of coal. In 2024, coal mining employed about 51,500 workers in Australia, or only about 0.36% of Australia's workforce (Australian Government, 2024). But over 90% of these workers were located in a small number of regions, including the Bowen Basin in Central Queensland, the Hunter Valley in New South Wales and the Latrobe Valley in Victoria (CPD, 2021; Kemp et al., 2021, p. 36). The coal industry is seen as economically significant both for Australia's national accounts and in key regions. As a result, Australia has had clear economic reasons not to embrace transition to lower carbon energy. But layered on top of this is the fact that regardless of its *actual* significance to Australia's economy, coal and the broader fossil fuel industry is widely *perceived* as crucially important, and the key fossil fuel-producing regions are also electorally significant at both the state and national levels. We turn now to the political drivers of Australia's inaction on energy transition.

The 'climate wars' and politicisation of climate change in Australia

Coal mining has enjoyed bipartisan support for decades, driven by a combination of the integral role of mining unions in the broader labour movement and the relationship between conservative governments and mining businesses (Baer, 2016; Edwards, 2019). But both ideological and economic considerations drove an intense politicisation of the public debate about climate change and energy policy from the mid-1990s onwards. This section gives a broad overview of the period, which is widely referred to in Australia as the 'climate wars'.

The 'climate wars' started with the LNP coalition government of conservative climate-sceptic Prime Minister John Howard. Howard combined fierce opposition to climate action and emission reduction measures with strong and unwavering support for the coal industry for over a decade in office from 1996 to 2007 (Crowley, 2021; Pearse, 2007, 2009). Howard's approach to climate and energy issues set the tone for at least the 15 years which followed his tenure. Climate and energy issues were at the heart of some of the most tumultuous years of federal politics in Australia's history, contributing to the rapid demise of a series of Prime Ministers.

Howard was voted out of government in 2007, losing his own seat in the process. He was replaced by Labor's Kevin Rudd who stood on a pro-climate agenda. Rudd implemented a range of policies aimed at reducing GHG emissions and accelerating investment in renewable energy. But the Rudd government's attempt to introduce an emissions trading scheme—the Carbon Pollution Reduction Scheme—was defeated in 2009 when it was blocked by the LNP opposition which retained a majority in the Senate (the upper house of Australia's bicameral parliament). Rudd was deposed from within his own party by Julia Gillard in 2011, who succeeded in implementing a national Carbon Pricing Mechanism as a key component of her government's Clean Energy Futures (CEF) Package (Hudson, 2019).

This progress was short-lived. Many CEF initiatives, including the carbon price, were rapidly reversed following the 2013 election of the LNP government led by Prime Minister Tony Abbott. Abbott replaced the CEF with a $2.55 billion Emissions Reduction Fund which paid businesses, local government and community organisations for emission reduction projects (Hudson, 2019). Amongst ongoing internal acrimony in Abbott's Coalition Government around climate and energy policy, Malcolm Turnbull deposed Abbott from office in September 2015 in a party-room split. He then attempted to implement a series of policy packages designed to create a more orderly transition to a low emissions energy system. These policy packages—which included a 'baseline and credit' scheme, a Clean Energy Target, and a National Energy Guarantee—were all rapidly undermined and defeated by LNP members firmly committed to maintaining the existing coal-based system (Hudson, 2019).

Turnbull was himself deposed by his treasurer Scott Morrison in a party-room split in August 2018, less than a year after Morrison had brandished a lump of coal in Parliament in a visual reminder that for him, sustaining Australia's coal industry was a higher priority than reducing carbon emissions. As Prime Minister, Morrison successfully mobilised concern about potential job losses in coal-dependent regions including Central Queensland, the Hunter Valley and the Latrobe Valley as part of a political strategy which helped him win the 2019 federal election, which he had been widely expected to lose (Hudson, 2019). During the campaign, the Labor opposition had proposed establishing a 'Just Transition Authority', and its election loss contributed to the sense that the electorate did not support ambitious climate change or energy transition policies (Edwards et al., 2022; see also Colvin, 2020).

The chaotic and acrimonious political environment at the federal level over more than 20 years set the tone for Australia's lack of engagement with the looming transition. Between 1996 and 2022, it was not politically safe to discuss an energy transition, let alone to plan for it. It is no exaggeration to say that 'climate change' became a dirty word, stripped out of the names of institutions and agencies across both Australia's federal government and many state governments. Particularly from 2010 onwards, a small number of prominent politicians supported by sections of the media successfully captured the public discourse to equate 'transition' with unemployment and an abandonment of regional communities (Edwards et al., 2022).

The 'climate wars' prevented Australia from developing a productive public conversation about climate change and energy transition and the implications of both for Australia. But it also led to the emergence of advocacy and activism in new places. The courts became key sites where climate-related arguments could be publicly pursued. With governments unwilling to entertain their arguments, Australian climate activists took them to the courts, employing a range of legal strategies aimed at highlighting broad legal and ethical responsibility for climate action as well as halting specific coal-mining projects and developments (Peel, 2023).

The febrile political climate also led to a growth in action in the finance sector, where shareholder activism became increasingly prominent. Led at first by established environmental groups such as Greenpeace, finance sector-focused groups such as Market Forces, the Australian Centre for Corporate Responsibility and the Investor Group on Climate Change increasingly took the lead. These groups drew on the broader international consensus that climate change was a key financial risk to target their actions in Australia. For instance, the growing international influence of the Task Force on Climate-Related Financial Disclosures (TCFD) was reflected in increasingly strong advice from financial regulators to Australian companies to align their climate disclosures with TCFD norms, even in the absence of it being a regulatory requirement.

Two prominent examples stand out. In 2019, the Reserve Bank of Australia observed that 'climate change is exposing financial institutions and the financial system more broadly to risks that will rise over time, if not addressed' (RBA, 2019, p. 1). In 2021, the Australian Prudential Regulation Authority advised companies that it 'anticipates the demand for reliable and timely climate risk disclosure will increase over time, and for institutions with international activities there is a need to be prepared to comply with mandatory climate risk disclosures in other jurisdictions' (APRA, 2021, p. 19).

But below these more visible legal and financial arenas, on-the-ground activism, campaigning and community organising continued during this period, albeit rarely couched explicitly in terms of climate change or energy transition. The case studies in this book are examples of how particular regional communities in fossil fuel-producing regions laid the ground for the sea change of 2022.

Local action began to penetrate the decision-making of state governments, particularly in recent years. States with limited fossil fuel resources and/or ready access to renewable energy predictably led the way. South Australia is notable in this regard. Fully reliant on coal in 2002, it was the first Australian state to achieve bipartisan political support for ambitious clean energy policies maximising the State's advantages in abundant solar and wind energy. Similar, if more muted, shifts were also visible in NSW, in which the coalition LNP government moved ahead of its federal counterpart by announcing goals of a 35% reduction in emissions by 2030 and net zero economy by 2050 in its 2020 *Net Zero Plan.* In 2022, the Government of Queensland, a state with a particularly strong history of political support for fossil fuel industries, launched the *Queensland Energy and Jobs Plan*, which includes a commitment to achieving 80% renewable energy by 2035.

The imperative of a 'just transition'

Paradoxically, Australia's prolonged period of political paralysis on climate and energy transition policy provides an emblematic example of the fact that justice is essential to acceptance of an energy transition as it is other transformations to sustainability (Martin et al., 2020).

In the last few years, justice has become increasingly prominent in climate change and energy debates through work on 'energy justice' and 'climate justice' (Edwards, 2021). Scholars and policymakers have both begun to re-engage with the idea of 'just transition' which first emerged in the North American labour movement in the 1970s and crystallised in the mid-1990s in the context of debates about how to move away from highly polluting and toxic industries such as logging and chemical production (Stevis and Felli, 2015; García-García et al., 2020; Stevis, 2023). Brian Kohler, a delegate from the Energy and Paperworker's Union of Canada, explained the concept in a speech to the Persistent Organic Pollutants Conference in Chicago that 'If society must make some tough choices about which economic activities we are willing to continue and which we are willing to forego, a structured transition or "just" transition program is necessary, if the costs of those decisions are to be shared fairly' (Kohler, 1996).

In 1997, the Oil, Chemical and Atomic Workers' Union of the United States became the first group to adopt a resolution specifically calling for a just transition (Stevis and Felli, 2015), but in recent years the energy transition has been the primary area that just transition ideas have been pursued (Harrahill and Douglas, 2019). In large part, this has been due to the work of the international labour movement, particularly the International Trade Union Confederation (ITUC) and International Labour Organization (ILO), which partnered with the UNEP to write the 2008 report *Green Jobs: Towards Decent Work in a Sustainable, Low-Carbon World* (UNEP et al., 2008).

Work on 'just transition' in the climate change context has tended to focus on the employment dimensions of the transition (e.g. ITUC, 2010; Olsen, 2010; ILO, 2015). For instance, the Paris Climate Agreement called for countries to take into account 'the imperatives of a Just Transition of the workforce and the creation of decent work and quality jobs in accordance with nationally defined development priorities' (UNFCCC, 2015). Since then, there has been growing acceptance that while a just transition must certainly provide for fossil fuel workers, the definition of 'just transition' must be considerably broadened to adequately address the climate change challenge (Wiseman and Wollersheim, 2021). As the then-ITUC General Secretary Sharon Burrow put it in 2017:

> the just transition will not happen by itself. It requires plans and policies. … transformation is not only about phasing out polluting sectors, it is also about new jobs, new industries, new skills, new investment and the opportunity to create a more equal and resilient economy. Social dialogue is the key.
>
> (Burrow, cited in Smith, 2017)

Academic work has observed that while the concept of just transition has considerable promise in bringing together labour and environmental concerns, it is nevertheless contested, with internal contradictions and tensions (Wang and Lo, 2021; Harry et al., 2024; see also Ciplet and Harrison, 2020; Bouzarovski, 2022; Stevis, 2023; Weller et al., 2024). This reflects ongoing debate about the core interpretations of just transitions, which Snell (2018, p. 552) argued 'tend to revolve around three key areas: (1) the balance between social and ecological "fairness"; (2) the role of the state and the type of state formation required to achieve JT; and (3) policy provisions and action'. One of the key outstanding issues is how to balance the imperative for a *fast* as well as a *just* transition (Ciplet and Harrison, 2020), because in Australia—as elsewhere—there is always the potential for the need for justice to delay necessary action (Weller, 2019; Harry et al., 2024).

The rest of this book

Changes in the relative prices of fossil fuels and renewable energy, state government policy leadership and legal and financial advocacy strategies have all played roles in creating a more favourable context for regional energy transition in Australia. The five case studies explored in this book provide important insights into the role of local factors and decisions in enabling and accelerating regional just transition policies and strategies. Indeed, one of the common conclusions from our case study authors is the importance of understanding and respecting the unique histories, cultures, challenges, concerns, knowledges and experiences of local communities.

Chapter 2 turns its attention to Port Augusta (South Australia), where a coalition of community groups and locals worked to stimulate new renewable investments to replace the brown coal generators which closed in 2016. Initial optimism about the potential for community-led change has been tempered by the failure of state and commonwealth governments to provide sustained, proactive planning, coordination and resourcing to support the energy transition.

Chapter 3 focuses on the Latrobe Valley (Victoria), which faced a rapid and unplanned transition away from coal-fired power generation starting in early 2017, followed by several attempts from local stakeholders and the state government to initiate transition planning for the remaining coal plants and mines in the valley. While the sudden 2017 closure of the Hazelwood coal-fired power station created massive local challenges it also triggered a rapid, concerted response from communities, trade unions and local and state governments to identify and support the growth of new industries, services and employment opportunities.

Chapter 4 focuses on Collie (Western Australia), where the state government-owned electricity generator has been working with community groups to structure the move away from coal-fired power generation in response to climate change. While all five case studies highlight the importance of understanding the ongoing legacies of colonial exploitation of First Nations lands and resources, Collie provides a particularly powerful example of the importance of inclusive participation of all segments of community led by First Nations Elders.

Chapter 5 examines the Hunter Valley (NSW), where coal-fired power stations are closing down even as export-oriented coal mining has boomed and a coalition of community groups, unions and environmental groups have put transition on the agenda. Significant challenges remain, however, in identifying alternatives to the still-pervasive coal industry.

Chapter 6 explores the Gladstone region (Queensland), where the community and local government have been attempting to put in place mechanisms to ensure a structured and fair transition from gas and coal exports by encouraging new investments in renewable industries. While the energy transition task in Gladstone has been particularly challenging, the success of recent energy transition dialogues between local communities, workers, business and governments illustrates the importance of creating inclusive and locally driven engagements.

These empirical chapters provide the basis for our concluding chapter (Chapter 7), in which we draw out the lessons which the case studies collectively teach about how to achieve rapid and just energy transitions.

The case studies explored in *Regional Energy Transitions in Australia* contain a number of key lessons. Firstly, it is an ethical, practical and political imperative that no workers and no communities are left behind in the transition. Justice, respect and inclusion provide essential foundations for public support of energy transition plans and strategies, particularly for the regional communities that are most directly affected. Secondly, all energy transitions are, in the end, local. As such they are shaped by the unique histories, cultures, challenges, concerns, knowledges and experiences of the local communities. The case studies highlight that for regional energy transitions to be just, they must provide for the workers who will be most directly affected. This confirms other research, but the case studies give hints to its form: well-planned, adequately funded re-employment, retraining and early retirement programmes for those working in the fossil fuel industries. But attention to workers alone will not provide a just transition, which must also deliver new high-quality jobs, economic renewal and environmental regeneration which benefits all members of the community, including those who are marginalised and whose histories have been punctuated with compounding disadvantage.

The single most significant lesson from these regions is that it is possible to put transition on the agenda even when politically and practically it seems impossible. Rapid, sustainable and equitable energy transitions are possible only when local communities are respectfully and inclusively engaged in the conversation and take ownership of the process for themselves. This requires local leadership, consensus building amongst actors with different primary motivators—from industry leaders to environmental and social groups—and above all the willingness to have respectful but frank conversations about what are often uncomfortable realities.

Notes

1 https://climatenetwork.org/resource/colossal-fossil-of-the-day-12-november-2021-australia/

2 There are many similarities between the specific transition in the Latrobe Valley and other forms of transitions, such as the closure of car manufacturing plants in South Australia.

3 See www.pm.gov.au/media/national-net-zero-authority and www.pm.gov.au/media/appointment-net-zero-economy-agency-and-advisory-board

4 https://budget.gov.au/content/03-future-made.htm?>

References

AEMO (2024) *Draft integrated system plan*. https://aemo.com.au/en/energy-systems/major-publications/integrated-system-plan-isp/2024-integrated-system-plan-isp?> (accessed 17 June 2024).

APRA (2021) *Prudential practice guide: CPG 229 climate change financial risks*. Australian Prudential Regulation Authority. www.apra.gov.au/sites/default/files/2021-11/Final%20Prudential%20Practice%20Guide%20CPG%20229%20Climate%20Change%20Financial%20Risks.pdf (accessed 21 June 2024).

Australia Institute (2022) *Climate of the Nation 2022*. The Australia Institute. https://australiainstitute.org.au/report/climate-of-the-nation-2022/ (accessed 21 June 2024).

Australian Government (2024) *Mining*. www.jobsandskills.gov.au/data/labour-market-insights/industries/mining (accessed 21 June 2024).

Baer HA (2016) The nexus of the coal industry and the state in Australia: Historical dimensions and contemporary challenges. *Energy Policy* 99: 194–202.

Bouzarovski S (2022) Just transitions: A political ecology critique. *Antipode* 54(4): 1003–1020.

Burke PJ, Best R and Jotzo F (2019) Closures of coal-fired power stations in Australia: local unemployment effects. *Australian Journal of Agricultural and Resource Economics* 63(1): 142–165.

BZE (2021) *Export powerhouse, Australia's $333 billion opportunity*. Beyond Zero Emissions. https://bze.org.au/research_release/export-powerhouse (accessed 21 June 2024).

Cahill A (2022) *What regions need on the path to net zero emissions*. The Next Economy.

Carbon Brief (2022) *Australia's biggest coal-fired power plant to shut years ahead of schedule*. www.carbonbrief.org/daily-brief/australias-biggest-coal-fired-power-plant-to-shut-years-ahead-of-schedule/ (accessed 21 June 2024).

Christoff P (2022) Mining a fractured landscape: The political economy of coal in Australia. In: M Jacob and J Steckel, eds., *The political economy of coal: Obstacles to clean energy transitions*. London: Routledge. pp. 233–257.

Ciplet D and Harrison JL (2020) Transition tensions: Mapping conflicts in movements for a just and sustainable transition. *Environmental Politics* 29(3): 435–456.

Clean Energy Council (2022) *Clean energy Australia report*. Clean Energy Council. https://assets.cleanenergycouncil.org.au/documents/resources/reports/clean-energy-australia/clean-energy-australia-report-2022.pdf (accessed 21 June 2024).

Climate Action Tracker (2023) *Temperatures*. https://climateactiontracker.org/global/temperatures/ (accessed 27 September 2024).

Climate Action Tracker (2024) *Net zero targets*. https://climateactiontracker.org/methodology/net-zero-targets/ (accessed 17 June 2024).

Colvin RM (2020) Social identity in the energy transition: An analysis of the 'Stop Adani Convoy' to explore social-political conflict in Australia. *Energy Research & Social Science* 66: 101492.

CPD (2021) *Who's buying? The impact of global decarbonisation on Australia's regions*. Centre for Policy Development. https://cpd.org.au/work/whos-buying-impact-decarbonisation-regoins/ (accessed 21 June 2024).

Crowley K (2021) Climate wars, carbon taxes and toppled leaders: The 30 year history of Australia's climate response. *The Conversation*. https://theconversation.com/climate-wars-carbon-taxes-and-toppled-leaders-the-30-year-history-of-australias-climate-response-in-brief-169545 (accessed 17 June 2024).

CSIRO (2024) *GenCost: Cost of Building Australia's Future Electricity Needs*. www.csiro.au/en/research/technology-space/energy/GenCost (accessed 21 June 2024).

Denis-Ryan A and Morrison K (2023) Australia can and should eradicate its gas supply gap – but not with more gas. *Renew Economy*, 3 April 2023. https://reneweconomy.com.au/australia-can-and-should-eradicate-its-gas-supply-gap-but-not-with-more-gas/ (accessed 8 October 2024).

Edwards GAS (2019) Coal and climate change. *Wiley Interdisciplinary Reviews: Climate Change* 10(5): e607.

Edwards GAS (2021) Climate justice. In: B Coolsaet, ed., *Environmental justice: Key issues*. London & New York: Routledge.

Edwards GAS, Hanmer C, Park S, MacNeil R, Bojovic M, Kucic-Riker J, Musil D and Viney G (2022) *Towards a just transition from coal in Australia?* London: The British Academy. https://doi.org/10.5871/just-transitions-a-p/G-E.

García-García P, Carpintero Ó and Buendía L (2020) Just energy transitions to low carbon economies: A review of the concept and its effects on labour and income. *Energy Research & Social Science* 70: 101664.

Geoscience Australia (2022) *Gas*. www.ga.gov.au/digital-publication/aecr2022/gas (accessed 17 June 2024).

Geoscience Australia (2024) *Coal*. www.ga.gov.au/education/minerals-energy/australian-energy-facts/coal (accessed 17 June 2024).

Groesbeck JG and Pearce JM (2018) Coal with carbon capture and sequestration is not as land use efficient as solar photovoltaic technology for climate neutral electricity production. *Scientific Reports* 8(1): 13476.

Harrahill K and Douglas O (2019) Framework development for 'just transition' in coal producing jurisdictions. *Energy Policy* 134: 110990.

Harry SJ, Maltby T and Szulecki K (2024) Contesting just transitions: Climate delay and the contradictions of labour environmentalism. *Political Geography* 112: 103114.

Hudson M (2019). 'A form of madness': Australian climate and energy policies 2009–2018. *Environmental Politics* 28(3): 583–589.

IEA (2022) *Overview: World Energy Employment 2022*. www.iea.org/reports/world-energy-employment/overview (accessed 21 June 2024).

IEA (2023) *Coal market update – July 2023*. International Energy Agency. www.iea.org/reports/coal-market-update-july-2023 (accessed 4 August 2023).

IEA (2024a) *Strategies for Affordable and Fair Clean Energy Transitions*. www.iea.org/reports/strategies-for-affordable-and-fair-clean-energy-transitions (accessed 21 June 2024).

IEA (2024b) *Renewables 2023*. Paris: IEA. www.iea.org/reports/renewables-2023 (accessed 17 June 2024).

ILO (2015) *Guidelines for a just transition towards environmentally sustainable economies and societies for all.* Geneva: International Labour Organization.

IPCC (2023) *Sixth Assessment Report, Climate Change.* www.ipcc.ch/report/sixth-assessment-report-cycle/

IRENA (2023a) *Renewables Jobs Nearly Doubled in Past Decade, Soared to 13.7 Million in 2022.* www.irena.org/News/pressreleases/2023/Sep/Renewables-Jobs-Nearly-Doubled-in-Past-Decade-Soared-to-13-Point-7-Million-in-2022?> (accessed 21 June 2024).

IRENA (2023b) *Accelerated Energy Transition Can Add 40 Million Energy Sector Jobs by 2050.* www.irena.org/News/pressreleases/2023/Nov/Accelerated-Energy-Transition-Can-Add-40-million-Energy-Sector-Jobs-by-2050?> (accessed 21 June 2024).

ITUC (2010) *Resolution on combating climate change through sustainable development and just transition.* Resolution 10, 2nd World Congress, International Trade Union Confederation, 21–25 June, Vancouver, Canada. www.ituc-csi.org/IMG/pdf/2CO_10_Sustainable_development_and_Climate_Change_03-10-2.pdf (accessed 5 November 2021).

Jakob M and Steckel JC (eds.) (2022) *The political economy of coal: Obstacles to clean energy transitions.* London: Routledge.

Kemp J, McCowage M and Wang F (2021). *Towards net zero implications for Australia of energy policies in East Asia.* Reserve Bank of Australia. www.rba.gov.au/publications/bulletin/2021/sep/towards-net-zero-implications-for-australia-of-energy-policies-in-east-asia.html (accessed 21 June 2024).

Kohler B (1996) *Sustainable Development: A Labor View.* www.sdearthtimes.com/et0597/et0597s4.html (accessed 5 November 2021).

Martin A, Armijos MT, Coolsaet B, Dawson N, Edwards GAS, Few R, Gross-Camp N, Rodriguez I, Schroeder H, Tebboth MGL and White CS (2020) Environmental justice and transformations to sustainability. *Environment: Science and Policy for Sustainable Development* 62(6): 19–30.

McGlade C and Ekins P (2015) The geographical distribution of fossil fuels unused when limiting global warming to 2°C. *Nature* 517(7533): 187–190.

Ogge M (2022) *War gains: LNG windfall profits 2022.* The Australia Institute. https://australiainstitute.org.au/wp-content/uploads/2022/10/P1289-War-gains-LNG-windfall-profits-2022-Web.pdf (accessed 18 June 2024).

Olsen L (2010) Supporting a just transition: The role of international labour standards. *International Journal of Labour Research* 2(2): 293–318.

Pearse G (2007) *High and dry.* Camberwell: Penguin Group Australia.

Pearse G (2009) *Quarterly Essay 33. Quarry vision: Coal, climate change and the end of the resources boom.* Melbourne: Black Inc.

Peel J (2023) Climate litigation is on the rise around the world and Australia is at the head of the pack. *The Conversation.* https://theconversation.com/climate-litigation-is-on-the-rise-around-the-world-and-australia-is-at-the-head-of-the-pack-210375 (accessed 17 June 2024).

Pew Research Centre (2021) In response to climate change, citizens in advanced economies are willing to alter how they live and work. www.pewresearch.org/global/2021/09/14/in-response-to-climate-change-citizens-in-advanced-economies-are-willing-to-alter-how-they-live-and-work/ (accessed 26 July 2024).

RBA (2019) *Financial stability review – October 2019.* Reserve Bank of Australia. www.rba.gov.au/publications/fsr/2019/oct/box-c-financial-stability-risks-from-climate-change.html (accessed 21 June 2024).

Readfearn G (2022) Australia's greenhouse pollution from coal higher per person than any other developed country, data shows. *The Guardian*, 20 May. www.theguardian.com/environment/2022/may/20/australias-greenhouse-pollution-from-coal-higher-per-person-than-any-other-developed-country-data-shows (accessed 21 June 2024).

SBS (2021) Business Council of Australia backs 2030 climate target. *SBS News.* www.sbs.com.au/news/article/business-council-of-australia-backs-2030-climate-target/evsbti0r4

Smith S (2017) *Just transition: A report for the OECD.* www.oecd.org/environment/cc/g20-climate/collapsecontents/Just-Transition-Centre-report-just-transition.pdf (accessed 17 June 2024).

Snell D (2018) 'Just transition'? Conceptual challenges meet stark reality in a 'transitioning' coal region in Australia. *Globalizations* 15(4): 550–564.

Stevis D (2023) *Just transitions: Promise and contestation.* 1st ed. Cambridge: Cambridge University Press. www.cambridge.org/core/product/identifier/9781108936569/type/element (accessed 29 April 2023).

Stevis D and Felli R (2015) Global labour unions and just transition to a green economy. *International Environmental Agreements: Politics, Law and Economics* 15(1): 29–43.

UNEP, ILO, IOE and ITUC (2008) *Green jobs: Towards decent work in a sustainable, low-carbon world.* Report. www.ilo.org/global/topics/green-jobs/publications/WCMS_158727/lang--en/index.htm (accessed 2 November 2021).

UNFCCC (2015) *Paris climate agreement.* https://unfccc.int/sites/default/files/english_paris_agreement.pdf (accessed 8 October 2024).

Wang X and Lo K (2021) Just transition: A conceptual review. *Energy Research & Social Science* 82: 102291.

Weller SA (2019) Just transition? Strategic framing and the challenges facing coal dependent communities. *Environment and Planning C: Politics and Space* 37(2): 298–316.

Weller S, Beer A and Porter J (2024) Place-based just transition: Domains, components and costs. *Contemporary Social Science 19*(1–3): 355–374. https://doi.org/10.1080/21582041.2024.2333272

Wiseman J. and Wollersheim L (2021) *Building just and resilient zero carbon regions*. Melbourne Climate Futures. www.unimelb.edu.au/__data/assets/pdf_file/0009/3934404/Wiseman-and-Wollersheim,-2021_MCF-Discussion-Paper_final.pdf (accessed 17 June 2024).

2 Energy transition in Port Augusta, South Australia

Lisa Lumsden and Linda Connor

Introduction

Port Augusta is further down the fossil fuels to renewables 'energy transition' road than most of Australia's coal regions. In just nine years from 2015 to 2024, the region has gone from a 60-year history of operating South Australia's entire coal electricity generation capacity (784 MW) to having no coal industry at all. Instead, as of 2024 it is hosting 843 MW of wind and solar generation capacity with 256 MW more in construction. This is all part of a larger South Australian energy transition that is on track to become 'the first non-hydro grid in the world to reach 100 per cent renewables' by 2027 (Parkinson, 2024).

The local government area of Port Augusta is located at the inland tip of the Spencer Gulf in South Australia, bounded by sea, semi-arid lands and the iconic Flinders Ranges. Port Augusta's population of around 13,500 falls into the lowest category of socio-economic disadvantage rating in Australia (ABS, 2021b). Aboriginal and Torres Strait Islanders make up 20.4% of this population, well above the national average of 3.2% (ABS, 2021a). There are two native title groups, Nukunu and Barngarla, that in recent years have won native title determinations in Port Augusta and adjacent lands. The majority of the First Nations population in the area however are not from these groups, they are long-term residents who were displaced from their traditional lands in past years as well as more transitory Aboriginal residents from remote communities.

DOI: 10.4324/9781003585343-2

Early change

The first signs of energy transition in the Port Augusta area came in 2004 when Lincoln Gap Wind Farm was proposed on part of a large pastoral property 15 km west of the town. In 2006, the developers secured planning approval for a 126 MW project, making the front page of the local paper. This decision planted the seed that renewable energy might be a new economic opportunity for Port Augusta, but for many years nothing happened to progress the wind farm.

Then in 2012 the Gillard federal Labor government proposed providing payments to coal generators to close as part of a renewed commitment to stronger climate action (Cullen, 2012, see Chapter 1). Since the Port Augusta power stations were burning some of the dirtiest brown coal in Australia, they were near the top of the closure list. Ultimately, the payment for closure initiative became a casualty of the 'climate wars' discussed in Chapter 1, but despite this the 400 regional power station and coal mine workers, as well as contractors, those in local supply chains, Port Augusta Council and residents became increasingly aware of the possibility of closure.

Historical context

Port Augusta townspeople were familiar with community tensions associated with the power stations. On one hand, their establishment by the state government in the 1950s and 1960s had made a long-term contribution to the local economy. On the other hand, their chimneys distributed contaminated ash all over the city, settling in houses and on rooves, polluting both their air and the drinking water that was collected in rainwater tanks from those rooves. The health problems associated with this pollution sparked community anger that peaked in the early 1980s with a backlash, championed by then Councillor Joy Baluch, that forced the state government to install precipitators (filters) on the chimney stacks of the Playford A and B power stations.

In 1998, the Port Augusta power stations were privatised and a year later so were the railways. These were the two major employers in town and the combination resulted in a kind of community collapse. As one resident remembers: 'every second house was up for sale'. In the subsequent years the power stations had various owners, until they were ultimately acquired by Alinta Energy in 2006.

The workers always knew the power stations couldn't last forever, either because the coal would 'run out' or because the stations would reach their extended end-of-life. Playford A had been decommissioned in 1985. But growing support for climate action combined with increasingly cheap renewable energy applied a new and more immediate pressure on the local industry.

Opportunity and cracks

In late 2011, the independent renewable energy think tank Beyond Zero Emissions (BZE) presented the Port Augusta Council with a 'Zero Carbon Australia Stationary Energy Plan'. The Plan positioned Port Augusta at the centre of an interconnected national network of concentrating solar thermal (CST) and wind generation (Wright and Hearps, 2011). CST is a renewable technology that involves directing the sun's heat via mirrors into a receiver and storing that heat in tanks for dispatchable thermal power generation. The BZE proposal immediately received enthusiastic support from the Council, led by Mayor Joy Baluch.

In February 2012, Alinta announced it was putting Playford B Power station into 'reserve storage', leaving Northern Power Station as the last operating coal generator in South Australia.

Community activation

In March 2012, after a well-attended town hall meeting about the prospects of CST, a small group of locals came together facilitated by environment groups to consider what actions they could take to transition from coal to solar thermal. This group became known as the Repower Port Augusta (RPA) local group. They were a volunteer committee, averaging around 12 members, including a power station worker, a City Councillor (one of the authors of this chapter), a nun, teachers, health workers, business owners, an administrative officer and retirees. Ages ranged from the mid-20s through to mid-70s with motivations ranging from job security, health, 'community good', social justice, opportunities for youth and climate change action.

The RPA local group saw CST as a cutting-edge technology that presented the town with a new image and reason for pride in a decarbonised future. It made sense to locals, who were already 'energy literate', because CST provided both energy storage and dispatchable power and could free Port Augusta residents from the 'employment versus health' trade-off.

With a $5000 grant from Council and strategic support from professional campaign manager Daniel Spencer, who was funded by the Australian Youth Climate Coalition (AYCC), the RPA group conducted a 'community poll' asking Port Augusta residents whether they wanted the coal power station replaced with gas or CST. CST achieved almost unanimous support from the 3500 respondents and captured state media attention (ABC News, 2012). Via this poll residents asserted their stake in the energy transition, making it clear to both Alinta Energy and the government that they aspired for a green and prosperous future.

Next, a formal 'Repower Port Augusta Alliance' was established involving local government, business, unions, health and environment organisations. This alliance provided a network of influence, resources and campaign legitimacy.

In August 2012, the South Australian Energy Minster announced a trip to Nevada to visit an operational solar thermal plant. In September the same year, the AYCC organised a 'Walk for Solar', in which 30 people from across Australia walked 300 km from Port Augusta to Adelaide to show their support for a CST transition. The walkers were warmly received by a large rally on the steps of Parliament House, where both the State Premier and Port Augusta Mayor Joy Baluch spoke in support of the idea, receiving state-wide television news coverage.

BZE then released a detailed plan for 'Repowering Port Augusta' that proposed replacing the capacity of Port Augusta's coal power stations with a phased installation of six CST tower generators supplemented with wind turbines (BZE, 2012). The report highlighted Port Augusta's exceptional energy assets, notably the abundance of solar and wind resources combined with vast expanses of land and proximity to pre-existing grid connections due to its history of coal-fired power generation.

Soon after, Port Augusta's State Member of Parliament, Dan van Holst Pellekaan, secured bipartisan support for a Select Committee to investigate the effects of replacing the Port Augusta coal-fired power stations with a concentrated solar thermal power station when the coal supply from the Leigh Creek feeder mine was no longer viable.

This catalysed five years of 'Repower Port Augusta' campaign activities including an extensive program of lobbying, petitions, letter writing, submissions, newspaper articles, speeches, media interviews and meetings with key politicians, including Prime Minister Malcolm Turnbull and Premier Jay Weatherill.

Large-scale renewable energy interest

The RPA campaign presented Port Augusta as an unusually receptive and suitable community for renewable energy development. During 2013 and 2014, a number of international CST companies began to show interest in Port Augusta, making ambitious media announcements and vying to secure grid-accessible land. Supporting these initiatives was the South Australian Labor government, led by Premier Jay Weatherill, which benefitted from broad public support for actively pursuing an ambitious renewable energy development agenda. Ultimately, the US-based company Solar Reserve secured the RPA campaign's support for their 150 MW Aurora proposal on a pastoral property 30 km north of the town.

National policy uncertainty

During this period of economic opportunity in Port Augusta, the conservative Liberal National Coalition regained government at the Commonwealth level, and took a particularly antagonistic stance towards renewable energy, actively working to maintain a fossil-fuelled economy (see Chapter 1). This directly opposed the South Australian government's policy and affected investor confidence in renewables, with the effect that the transition to renewable energy in Port Augusta stalled.

Closure shock

In late 2014, Alinta Energy management told workers it was reviewing its Port Augusta operations. In an April 2015 presentation to the Port Augusta Council, Alinta's CEO stated that it was exploring new coal seam deposits and planned to operate the Northern power station until 2030–2032. But just two months later, the company announced it would close all its mining and generation operations 'within three years', stating the business had become 'increasingly uneconomic' (ABC News, 2015). At the time about 200 people worked at the power station in Port Augusta and a further 200 at the mine in Leigh Creek. Workers, contractors, families, residents, businesses, Council and even the local Member of Parliament woke up to learn the news via the media. Shock and despair rippled through the community.

As things played out the closures came even sooner than anticipated: the Leigh Creek mine was closed in nine months, and the power station was closed within the year. A few years later it was revealed that Alinta had sought payment from the State Government to stay open longer, but negotiations had failed.

Crisis and 'the fork in the road'

In September 2016, a cyclone damaged the state's major transmission lines, triggering a cascade of power system failures. South Australia's entire electricity distribution network shut down for several hours, and some regional areas were left without power for days. The economic impact of the blackout, known as the 'Black System Event', was estimated to be $367 million (Harmsen, 2016). Without Port Augusta's 520 MW of capacity the state electricity grid was also vulnerable. This became evident during the 2016/17 summer heatwaves when the Australian Energy Market Operator (AEMO) ordered rotational load shedding (planned blackouts in designated areas) 'due to a lack of available generation supply in SA' (Harvy and Shepherd, 2017). The cyclone and the rotational load shedding were coupled with some of the country's highest electricity prices, and the South Australian public outcry grew fierce.

The conservative federal government used the South Australian problems to attack the viability of renewable energy and criticise the state Labor government's ambitious pro-renewables agenda. Premier Weatherill retaliated by berating the Federal Energy Minister at a press conference and reiterated his commitment to renewables as both an economic opportunity and climate responsibility.

Under sustained political pressure the State Government could have wound back their renewables ambitions, instead they developed a suite of management strategies. One involved Elon Musk, who proposed (via Twitter) to build a 100 MW Tesla battery facility in less than 100 days. The battery, which was the world's largest at the time, became operational on schedule in December 2017 (AAP, 2017).

The 'renewables rush'

The need for rotational load shedding highlighted the gap in South Australia's energy market left by the closure of Port Augusta's coal-fired power stations. Port Augusta was perfectly positioned to host new energy developments, with underutilised transmission capacity, high quality solar and wind resources and a community receptive to large-scale renewables.

Between 2016 and 2019, many large-scale renewables projects were announced for the Upper Spencer Gulf, always promising hundreds of jobs. These projects included wind, solar photovoltaics, solar thermal (graphite, water and molten salt), pumped hydro (salt and mains water), 'big batteries' and hybrid projects where a mix of renewables are integrated into one project. It felt like a 'renewables rush' and local hopes for a big new renewables industry began to grow, with the giant CST Aurora project the jewel in the crown. Renewables seemed like an unmissable chance to kickstart the regional economy, lift community morale and reframe Port Augusta's identity, from a dirty coal industry town to a modern town at the forefront of the national energy revolution.

A people-powered win

In 2017, the State Government opened a major tender process calling for applications to supply the government's power needs. In response, the US company Solar Reserve developed their Aurora CST project into a competitive tender bid. The bid was given an extra boost when the Centre Alliance Party, led by Federal Senator Nick Xenophon, negotiated $110 million in concessional Commonwealth finance to support the Aurora project (Macdonald-Smith and Evans, 2017).

The RPA campaign manager Daniel Spencer led the RPA Alliance to ramp up its efforts for a tender win. A petition was circulated, attracting signatures from across Australia, and a crowdfunded billboard was erected facing Parliament House in Adelaide, where another well-attended rally took place.

This period was a difficult campaign sprint for the local RPA group behind the scenes. Volunteers were tired and had other life pressures. The group Chair had lost his job with the power station closure and hadn't yet found new work. Another member was hospitalised with mental health problems following her husband's redundancy at the coal mine. Had it not been for the funded campaign manager, the momentum would have been lost.

Then in August 2017 the Weatherill Government announced it had awarded Solar Reserve a 20-year contract for power, sourced from the proposed Aurora CST project (The Guardian, 2017). The development application for the project was lodged the next month and approved in January 2018.

The local RPA group was elated. The community felt vindicated and there was a sense of social and regional justice. The campaign had delivered a win that demonstrated the power of self-determination and democracy. Despite this and other efforts to restore public confidence, in 2018 Labor lost the State election, and perceptions of poor energy management were cited as a major reason. The incoming conservative Liberal government appointed Port Augusta's MP Dan van Holst Pellekaan as Energy Minister, and continued the state renewables transition, albeit in what was described as a 'more considered and controlled manner'. A post-COVID state election in 2022 then saw the return of the Labor government.

'Running the gauntlet'

Port Augusta's experience revealed that the transition to renewable energy is like running a gauntlet. Proposed projects are held back and often 'slayed down' by technological capacity, location and land limitations, stakeholder objections to planning applications, changes to government policy, ownership churn, grid connection approvals, power purchase negotiations, and finance challenges. Only a few make it across the line.

Over time, Port Augusta's potential for energy transition has translated into reality, if not on the scale or in the way originally hoped for. At the time of writing in 2024, there are four operating renewable generators in the local government area and its nearby hinterland.

The first is Sundrop Farm on the outskirts of Port Augusta town. Sundrop Farm is a four-hectare hydroponic tomato farm that desalinates water from the Spencer Gulf. Opened in 2017 and currently owned by Centuria, it includes a 39 MW CST mirror array with a 110 m tall receiving tower that glows like 'a second sun', supplemented by solar photovoltaics.

The second is Bungala Solar Farm owned by Italian company Enel Green Power. Bungala is a 275 MW solar photovoltaic project which occupies 800 ha of semi-arid pastoral land leased from the Bungala Aboriginal Corporation (BAC) north-east of Port Augusta. The land had previously been awarded to BAC by the state government as freehold title, despite Nukunu and Barngarla groups each claiming native title rights. The solar farm began construction in 2016, with Stage 1 becoming operational in 2018 and Stage 2 in 2020. At that time, it was the largest solar farm operating in Australia.

The third is Lincoln Gap Wind Farm, owned by Nexif Ratch Energy. Following the power station closure this hitherto stalled project commenced Stage 1 construction in 2017 and become operational in 2020. Since then, Stage 2 has been completed and Stage 3 is approved and pending construction. When complete, Lincoln Gap Wind Farm will be 468 MW.

The fourth is the Port Augusta Renewable Energy Park, a 210 MW wind and 107 MW solar photovoltaic hybrid project located south-east of Port Augusta town. Developed by Irish company DP Energy then sold to Spanish transnational Iberdrola in 2016, it was the only Port Augusta project that faced any real public resistance. Resistance centred on the wind turbine heights and visual impacts on the Flinders Ranges landscapes, though some native title claimants also held significant concerns which did not become public. It secured planning approval and began construction in 2019, commencing operation in 2022.

The Aurora CST project, however, did not survive the gauntlet. Despite community support, a $110 million concessional finance loan, a 20-year government contract and deadline extensions, Solar Reserve was unable to secure project finance. In April 2019, the project was cancelled. The host landowner and other residents were left to hear the news in a media announcement from the State Energy Minister (ABC News, 2019). This failure showed the Port Augusta community that it hadn't fully comprehended the intricacies of 'the gauntlet' nor the inexorable momentum of capital accumulation. After this experience there was no misunderstanding: the announcements mean nothing 'until there's a shovel in the ground' and renewable energy companies 'are here for the money, not us'. From that point onwards energy transition happened *to* the Port Augusta community, not *with* it.

'Cleaning up' after coal

After the closure of Northern power station Alinta 'sold' the site to a new internal subsidiary company, called Flinders Power. Effectively, this shifted the remediation responsibility to a smaller company, that if pushed could declare bankruptcy and walk away. This is significant because over the 60 years of coal power generation the ash dam grew to an area of 273 ha and a depth of 12–15 m (Greenpeace, 2018). The power station closure meant that the water pumps that kept the ash wet, preventing it from becoming airborne, were turned off. In the hot, dry, summer conditions that followed, winds exposed residents to a protracted dust storm of dried-out ash. The State Environment Minister and the Environment Protection Agency (EPA) were slow to respond and residents took their outrage to the media. Eventually a management plan was initiated, but residents viewed the lack of urgent action to safeguard public health as offensive and negligent.

To remediate the ash dam the EPA required Flinders Power to cover it with 10–15 cm of soil. Over two years, topsoil was sourced from nearby, despite it being a special Nukunu cultural site. Giant sprinklers and air-dropped dust suppressant were used to help prevent the soil blowing away, and native seed was planted with mixed results to bind the topsoil together.

By contrast, the power station infrastructure removal process was fairly smooth. It involved salvage auctions, vast metal recycling and significant asbestos management and removal. In 2018, when the iconic 200 m tall Northern chimney stack was demolished, crowds of locals watched on, many in sadness or nostalgia. The chimney had defined the landscape for generations and was a symbol of the region's former identity and purpose.

Today the Leigh Creek coal mine township 300 km to the north of Port Augusta is a small remote service centre. The population shrank from 505 people in 2011 to 91 in 2021 (ABS, 2011, 2021c). After the mine closure the exposed coal was covered with 2 m of inert matter and compacted. A 'Leigh Creek Future Town Plan' continues to be developed by the state government. The remaining locals are focusing on tourism opportunities and a coal gasification project is being trialled.

The mirage of economic opportunity

After the 2016 closure, the Council CEO reported an estimated economic loss to the local economy of $240 million annually. Despite this, the SA government taskforce for transition planning met in Adelaide, 300 km away, focused on immediate worker transition issues and initially lacked any representation from the Port Augusta Council. By default, much of the responsibility for managing the economic and social fallout was left to the indebted Council administration, other already stretched local economic development and social support organisations, small business owners and the RPA local group, without any additional resources. When locals became aware of the hundreds of millions of dollars in state and federal resourcing for the Latrobe Valley transition in the wake of the Hazelwood power station closure in 2017 (see Chapter 3), the feeling of injustice was bitter.

In 2017, $20 million in Federal economic stimulus funding was allocated across the far north and west region of South Australia. The application period was short and required 'ready projects' with matched funding. Because of this, the Port Augusta Council was not in a position to apply.

In those initial post-closure years, without multi-level leadership, planning and support, the local economy was weak. The state government's contract with the Aurora CST project would have been like a state funded economic diversification project, but as explained earlier it didn't eventuate. Rather it was the state led expansion of the Port Augusta gaol and the power station remediation works led by Flinders Power which offered the economic bridge.

Subsequently, Port Augusta experienced several years of economic stimulus through an influx of construction workers and contractors working on multiple renewable energy projects and also the state and federal government funded Joy Baluch Bridge Duplication Project. These projects were estimated to have mobilised $1.7 billion in collective capital investment, but it is difficult to ascertain what proportion of this was actually spent in the town, region or even the state. As a result, their real value to the regional economy remains unknown.

The ongoing economic contribution being made by the local renewables industry to Port Augusta, now that it is in the operational phase, has so far been relatively small. As this reality became apparent, locals began hearing that the long-term economic opportunities from energy transition were actually in green hydrogen production and export commodities, renewable supply chain manufacturing, decarbonising existing export industries and co-locating energy intensive industries with cheap 'behind the meter' electricity supply, not the renewables themselves. Neighbouring Whyalla trialled the manufacture of wind towers without market success. Lacking economic coordination, policies and market incentives to drive these types of industries to the region, local economic opportunities have not yet materialised. Like a 'mirage of economic opportunity' the benefits of the renewable industry seem always beyond reach.

Multiple other major diversification projects have been proposed since the power station closure, none of which have yet come to fruition. 'Announcement fatigue' is how one local leader describes the situation. One current circular economy proposal seeks to mine the ash dam for the manufacture of 'green cement' using a 6 MW hydrogen electrolyser (Department of Energy and Mining, 2023), but it will need to be seen to be believed.

Just transition for power station workers?

When the power station and mine closure was announced there was considerable public outcry in the media, but in the end the nearly 400 workers left quietly. Long-term employees were supported by government backed redundancy arrangements that had been negotiated when the power station was privatised in the 1990s. Those employed before privatisation could move to another (often obscure or supernumerary) role within government or take a subsidised redundancy package. Those who had joined after privatisation were entitled to their contractual redundancy packages through Alinta only. Around $3 million was allocated by the State government to support the Port Augusta and Leigh Creek workers to move into other work.

When Sundrop Farm opened in 2017 it was seen by some outsiders as a form of 'just transition', but the reality was rather different. Around 15 former power station workers secured work operating the CST facilities, but most didn't apply for the far lower-paid and unskilled agricultural jobs, most of which were ultimately filled by overseas workers from Asia and the Pacific.

Overall, it is estimated that about a third of the power station workers took early retirement packages. Another third left town to work in power stations around the country. Some relocated with their families, but others took Fly-in-Fly-Out (FIFO) or Drive-in-Drive-Out (DIDO) jobs while their families stayed locally. The remaining third eventually found other regional work in mining, other manufacturing, local building or electrical trades. All this meant the post-closure unemployment figures never showed any major or prolonged fluctuations, but was it *really* a 'just' transition?

Social impacts, empowerment and community pride

After the closure and despite the renewables developments that followed there was a widespread sense of dejection and uncertainty as many people dealt with the psychological and practical fallout. There were relationship break-ups, children had difficulties at school, and there was at least one related suicide. Some who tried FIFO or DIDO arrangements couldn't stand the new separation and either relocated elsewhere or found local work. Birth rates dropped and mental health, domestic violence, drug and alcohol issues increased.

Once Port Augusta Renewable Energy Park (PAREP) planning applications became adversarial, and especially after the Aurora project cancellation, community members felt they did not have power to influence the course of change. Naysayers were vindicated, and the hopes for community participation in renewable energy rollout dissipated as transnational corporations, funded by shareholders and public subsidies, dominated the field. While alternative models of energy transition exist, including community participation and ownership (Cahill, 2018), they were sidelined early in South Australia.

In an already disadvantaged community like Port Augusta the marginalised are easily displaced. During the peak renewable energy and bridge construction periods house rentals became too high for some locals, leading to increases in couch surfing and homelessness.

Jobs and training in renewables

Renewable energy projects are often spoken about with the job numbers that are projected to be involved, as these form part of the social contract. But these figures are frequently misleading. For example, '300 jobs in construction' means a total of 300 full time equivalent Australian jobs will be supported over the whole construction period, which may be 1–3 years. It does not mean 300 new or local jobs.

The Port Augusta renewable energy projects employed hundreds, perhaps thousands of people, many for short periods, as required for different stages of the project. But most jobs were not local, since project design, planning and finance staff, as well as lead construction companies were usually based in Australian cities or overseas. On site, many construction employees have ongoing contracts with the major contractor or sub-contractor and have pre-existing renewables-specific experience, skills, and certification. These employees work to short timelines under huge pressure, as part of mobile crews that move on elsewhere when their phase of work is finished.

That said, contractors want to hire locals where practicable and access labour hire agencies and social networks to find people, sometimes poaching workers from neighbouring project teams. In Port Augusta, regional residents found employment as truck drivers, general and casual labourers, in administration, security and some electrical roles. A local labour hire officer described how he proactively aggregated work across projects to create longer-term contracts for people. Some companies made efforts to employ Aboriginal residents but there were no prescribed targets for their participation. All in all, for many locals, the lure of stable well-paid positions at the steelworks in Whyalla or at remote mines further north were far more appealing.

Local workforce capacity also limits participation, and this is exacerbated in lower socioeconomic zones. Flexibility to accommodate diversity requires resourcing, and long-term unemployed people require ongoing intensive support. Efforts to train up local people to participate during the construction phase were largely unsuccessful in Port Augusta, in part due to many years of TAFE system fragmentation, the lack of traineeship and apprenticeship offerings and the uncertain timelines of construction projects.

Taking a creative approach, the owner of Bungala Solar Farm worked in collaboration with Bungala Aboriginal Corporation to initiate a food catering training program for local Aboriginal residents, initially to feed construction workers. This has developed into a successful longer term catering enterprise, but there are few such examples. Recent negotiations with native title corporations are improving this situation for native title holders, if not for the wider Aboriginal population.

It is expected that the currently operational 840 MW large-scale wind and solar power generation around Port Augusta will collectively employ around 40 local people per year over the next 30 years. Some of this work is skilled electrical and maintenance roles but the majority is lower skilled, periodic work, such as solar panel cleaning and site maintenance. Senior company managers are in head offices elsewhere and highly skilled wind turbine maintenance crews tend to operate in a scheduled FIFO/DIDO fashion, coming to town for only a few weeks a year.

Council leadership, capacity and control

In 2011, the Port Augusta Council was quick to recognise the possibilities in the BZE report and the value of the RPA campaign. Advocacy from Mayor Joy Baluch gave locals early confidence. After her passing, Mayor Sam Johnson and CEO John Banks championed renewables with equal vigour and worked closely with the RPA Alliance and residents.

The legacy impacts of South Australian electricity legislation soon emerged as one of the Council's greatest challenges. Prior to the 1990s, when electricity was almost entirely owned by the State, generators were exempt from paying council rates and this exemption was maintained when the sector was privatised. Instead, the State government brokered a 'debenture agreement' between the Port Augusta power station owner and the Council, which was worth around $500,000 annually by 2016. The closure of the power station abruptly removed this revenue source, contributing to hikes in local Council rates which snowballed into community anger directed at the Councillors.

The prospects of multiple large-scale renewable energy developments offered new hope for supplementary rates income, but Council soon found they were unable to levy appropriate rates on the renewables projects due to the continuation of the legislated rates exemption. To date, despite coordinated lobbying by several regional councils, successive state governments have maintained the exemption and avoided new debenture arrangements. They have argued that charging rates would increase power prices and undermine the investment attractiveness of the state. In the meantime, the Port Augusta Council receives no additional income from hosting renewables, which council officers argue results in the already disadvantaged community essentially subsidising multi-national company profits.

Over many years, Council has complained that its authority has been diminished through centralised decision-making as well as shifts in unfunded, burdensome administrative requirements to it. Port Augusta Council has been distanced from large development approvals within their jurisdiction and are instead 'invited to respond', with limited expertise and capacity, to hundreds of pages of highly specialised renewable energy planning applications.

The Council also received the brunt of community queries and complaints related to renewables projects, without any additional administrative resourcing and minimal power to control outcomes. The depleted Council was responsible for working with the EPA or companies when there were issues, such as the dust management problems discussed earlier or when road damage became apparent. And the Sundrop Farm trial site has been left derelict and the Port Augusta Council is left to pursue the multi-national company that owns it for restitution.

Native title holders and First Nations people

Developers began circling Port Augusta before the National Native Title Tribunal determinations were handed down. This meant that in some cases developers undertaking the planning process entered negotiations with claimants *before* they had organised into the unified Prescribed Body Corporate required for native title governance, a decades-long process for Nukunu and Barngarla. As a result, a few companies took advantage, which some Traditional Owners refer to as 'corporate lowballing'. In some cases, the engagement practices of these companies have left lasting wounds within the title holder groups. Native title determinations have since been handed down (between 2016 and 2022), but all were *after* the energy proponents around Port Augusta had secured planning approval, leaving no clear processes for access and negotiation for benefit sharing.

Native title corporations now established in the Upper Spencer Gulf are supportive of renewable projects provided they are treated as negotiating equals, properly resourced and their rights respected. Better outcomes are now emerging, including equity sharing, good income for fair lease contracts, cultural monitoring work, training programs and employment opportunities. Sites of high cultural heritage and spiritual significance must be respected, or agreements may falter.

Notably however, those who are not native title holders have no access to the benefits subsequently negotiated through these native title determinations. As noted earlier in the chapter, this includes the majority of Aboriginal residents in Port Augusta.

Legacy benefits from energy transition

Decarbonising electricity generation is essential to arrest further climate change, so the closure of the power station and the roll out of renewables in Port Augusta leaves a lasting global benefit. Despite early crises, energy stability has now returned to South Australia and the wholesale price for electricity is consistently lower than Queensland and New South Wales. This is in large part because the South Australian electricity supply has been strengthened and over 74% of state generation is from lower cost renewables (OpenNEM, 2023).

In a neoliberal free market, aimed at producing energy at the cheapest cost for maximum profits, the strongest competitors are those who can secure projects at the largest scale. In Port Augusta all four renewable projects are owned by multinational corporations. They benefit from the pre-existing state-owned transmission grid and a government policy and planning environment designed to support their development.

Over the next 30 or more years, the investors in these companies will collectively benefit from billions of dollars in profits. But there will also be some local benefits. In Port Augusta, a few landholders negotiated far better land lease deals than others, but all appreciate the additional income they will receive for the life of the projects. The broader population benefits too from breathing clean air and avoiding coal-related illnesses.

Community benefit schemes are common within the renewables industry as they are seen to foster good will and social license. They include grant programs, partnerships, scholarships, training or opportunities for co-ownership. Best practice benefit sharing models incorporate aspects of local governance and involve aggregating and leveraging funds to achieve larger strategic impact and/or projects developed to respond to local needs (Hicks and Mallee, 2023).

In Australia, the financial value of community benefits from renewables falls in a median range of $500 to $800 per MW annually for wind and $500 to $1000 per MW for solar (Muralidharan, 2023). Given this, the projects in the Port Augusta area could be worth around $402,000 to $719,600 per year to the community. But in reality, the contributions and community benefits delivered by projects fall far below these industry standards, meaning at present, they contribute no obvious long-term community legacy.

This raises the question of whether being receptive to renewables development may have undermined Port Augusta's benefits. Had the community taken an antagonistic stance towards the industry, would they have secured a better legacy deal? Was this an opportunity missed, or can best practice benefits still be negotiated? What bargaining power do they have now?

Key learnings from the Port Augusta energy transition journey

Understanding the local context and history is crucial

Energy transition is an ongoing process shaped and influenced by the unique historical context of communities, towns and regions. Coal workers and communities need to feel their contributions to past economic growth are fully appreciated. Fossil fuel workers understand the climate change situation, but being proactive in renewable energy advocacy can feel risky.

Bold government policy and consistency matters

Steadfast governments willing to make bold decisions can change the course of history, despite paying a short-term electoral price. Uncertainty stifles progress. Communities and investors need clear and consistent State and Federal leadership to support an equitable renewable energy transition.

Preparation is key

Sudden closures of major industries in regional communities can undermine economic and social stability and exacerbate disadvantage. When energy issues become an electoral threat, government action can be swift. Early community planning and coordination can help to moderate shocks.

Broad community-led energy transition needs vision and support

Communities need tangible, shared and inspiring visions to become activated for change. When the vision speaks to the unique circumstances of a place, is compatible with local skills and identity, and fosters community pride, it has a better chance of securing support.

Communities can play a powerful role in attracting the industries they want to their region, including through committed volunteers, financial resourcing, proactive leadership, alliances of organisations and politicians who champion their goals. Well-coordinated community groups, even when small, can be active and influential players in determining how the energy transition plays out. It is however important to recognise that volunteer community campaigners get tired and need renewal, including support from funded and well-connected campaign managers.

'Just transition' means good jobs and real choices

A 'just transition' is not 'any job'. The type of replacement jobs, skill compatibility, quality, meaning and conditions are key. Early retirement can be a popular or desirable choice for older power station workers to deal with closures.

'After coal' land management requires strong leadership at multiple levels

State and Federal Governments persistently fail to take effective action to regulate corporate behaviour that seeks to minimise remediation responsibilities. In South Australia, the Adelaide based Environment Protection Authority is often poorly equipped to set high standards, monitor and manage issues and emergencies and enforce compliance in the regions.

Capturing economic opportunities after a shock requires resources

The social and psychological impacts of major economic readjustment require political acknowledgement, resources and strong policy initiatives. Small regional communities have limited political influence which impedes their ability to attract attention and support.

Economic recovery and diversification require multi-level focus, planning and coordination centred around local community strengths and aspirations. It also requires consistent public and private investment. Mining towns that exist because of coal can struggle to find an equal alternative economic purpose after closure.

The local economic value of renewables lies beyond the projects themselves

Long-term economic value for communities hosting renewables is more likely to be found by developing local, national and international renewable supply chain manufacturing, decarbonising existing export industries and co-locating energy and labour-intensive industries with cheap 'behind the meter' electricity supply.

Council capacity to lead, manage and derive value from renewables is central to success

Councils are at the front line in regional communities. When renewable energy companies aren't required to pay appropriate Council rates (or equivalent) the social licence of the industry is undermined. State and federal governments need to entrust councils with more autonomy and resource their capacity to respond to concerns and manage energy transition well. Councils also need negotiating power that encourages renewable companies to foster and maintain relationships with them throughout the project development, operational and remediation stages.

Construction work in renewable projects is harder for locals to access than it might seem

Announced job numbers in renewable projects reflect the aggregate number of full-time equivalent jobs a project will support in the national economy. However, they are not necessarily new jobs or local jobs.

Long-term skilled work in the renewable construction industry mostly involves years of specialist training, meaning large contractors tend to use FIFO models for construction work. The work available to locals living near a construction project is therefore mostly lower skilled and short term. Diverse groups and long-term unemployed locals face particular barriers to employment within a high pressure and profit driven employment environment. Systematic change, advanced planning, contractor coordination and practical wrap-around worker support is required to enable these groups to access employment in the renewable construction industry.

Good company relations with host communities are crucial for strong renewable development

Companies can best serve the future of the renewable industry by communicating honestly and genuinely partnering with the host community throughout the project lifecycle, not just the planning phase.

All levels of government need to facilitate enduring and strategic legacy benefits in host communities, rather than leaving each one to negotiate alone. Measurable improvements in socio-economic equity must be designed into energy transition implementation. Community co-ownership in renewable projects is one of many ways to generate legacy benefits. Benefits will not occur as an automatic 'trickle down' effect of market-driven economic stimulus.

Justice for First Nations communities should be a central component of 'just transition' policies and strategies

Despite their rights under the *Native Title Act*, native title groups can still be marginalised in consultation and legal contracts. Weak and unfair agreements with native title corporations create lasting community division and can lead to the breakdown of social licence for developers and operators. In Port Augusta, native title determinations have improved outcomes for specific native title holding groups negotiating agreements with developers, but these arrangements do not apply to the majority of Aboriginal residents, who due to historical movement are not eligible. A policy framework to facilitate training tied to employment targets for First Nations residents in host communities is needed to achieve more equitable outcomes.

Conclusion

Port Augusta has experienced both the end of coal and the rise of renewables: a journey of highs and lows. As a population facing some of the greatest socio-economic challenges in Australia, the early days of energy transition presented an opportunity for community-led change, targeted improvements and legacy impact. But despite community hope and positivity, so far the transition has made a minimal long-term contribution to material well-being in Port Augusta. Socio-economic indicators have not improved (Goodman et al., In press). This is largely due to the fact that successive governments have essentially left the delivery of the transition to the logic of the market. Had Port Augusta's transition been more proactively planned, resourced and coordinated, the local community might have found ways to leverage the opportunities for lasting change and a social equity agenda.

The Port Augusta story speaks to the range of social, cultural, economic, structural, political and environmental factors that make up the complexity that is 'energy transition'. As Abram and colleagues have discussed, conflict is part of energy transition, but 'transitionist imaginaries suggest a gentle, gradual, consensual change and tend to depoliticize its real implications' (2023). We hope the insights from Port Augusta will enable energy transitions elsewhere to unfold in ways that are carried out *with*, are kinder to and are more beneficial for the regional communities involved.

References

ABC News (2012) Port Augusta backs solar thermal power. *Online News*, 23 July.

ABC News (2015) Alinta Energy to close power stations at Port Augusta and coal mine at Leigh Creek. *Online News*, 11 June.

ABC News (2019) Port Augusta solar thermal power plant scrapped after failing to secure finance. *Online News*, 5 April.

Abram S, Waltorp K, Ortar N and Pink S (eds.) (2023) *Energy futures: Anthropocene challenges, emerging technologies and everyday life*. Berlin and Boston: De Gruyter.

ABS (2011) *Leigh Creek (Urban Centres and Localities), Leigh Creek (L), Census All persons.* QuickStats, Australian Bureau of Statistics.

ABS (2021a) *Port Augusta, 2021 Census Aboriginal and/or Torres Strait Islander people QuickStats (by Local Government Area), Port Augusta, Census Aboriginal and/or Torres Strait Islander people.* QuickStats, Australian Bureau of Statistics.

ABS (2021b) *Socio-Economic Indexes for Areas (SEIFA), Australia, 2021.* QuickStats, Australian Bureau of Statistics.

ABS (2021c) *Leigh Creek (Urban Centres and Localities), Leigh Creek (L), Census All persons.* QuickStats, Australian Bureau of Statistics.

Australian Associated Press (AAP) (2017) South Australia turns on Tesla's 100MW battery: 'History in the making'. *The Guardian*, 1 December.

BZE (Beyond Zero Emissions) (2012) *Repowering Port Augusta.* Beyond Zero Emissions.

Cahill A (2018) Economic diversity in the energy sector: Post-capitalism in the here and now? *Australian Quarterly* 89(2): 22–27.

Cullen S (2012) Government scraps plans to shut dirty power stations. *Online ABC News*, 5 September.

Department of Energy & Mining (2023) *Landmark agreement for green cement processing*. Energy & Mining.

Goodman J, Bryant G, Connor L, Ghosh D, Marshall JP, Morton T, Mueller K, Rosewarne S, Heikkinen R, Lumsden L, Pampus M and Pillai P (In press) *Decarbonising electricity: The promise of renewable energy regions*. Cambridge: Cambridge University Press.

Greenpeace (2018) *Done & dusted? Cleaning up coal ash in Port Augusta*. Ultimo NSW: Greenpeace Australia Pacific.

Harmsen N (2016) South Australian blackout costs business $367m, fears summer outages on way, lobby group says. *ABC Online News*, 9 December.

Harvy B and Shepherd T (2017) Rolling blackouts ordered as Adelaide swelters in heatwave. *The Advertiser*, 9 February.

Hicks J and Mallee K (2023) *Is regional benefit sharing the New Frontier for Australia's Renewable Energy Shift?* Community Power Agency.

Macdonald-Smith A and Evans S (2017) Solar thermal project gets leg-up to beat rivals. *Financial Review*, August 15.

Muralidharan P (2023) *Building stronger communities, how community benefit funds from renewable energy projects support local outcomes*. Re-Alliance.

OpenNEM (2023) *Graph: Energy GWh/year*, generation data by state, all years. OpenNEM. Available at https://explore.openelectricity.org.au/energy/nem/ (accessed 27 September 2024).

Parkinson G (2024) South Australia locks in federal funds to become first grid in world to reach 100 per cent net wind and solar. *Renew Economy*, 21 July.

The Guardian (2017) Port Augusta solar thermal plant to power South Australian government. *The Guardian Online*, 14 August.

Wright M and Hearps P (2011) *Zero carbon Australia stationary energy plan*. 2nd ed.. Melbourne: The University of Melbourne Energy Research Institute and Beyond Zero Emissions.

3 The transition of the Latrobe Valley, Victoria

Dan Musil and Elianor Gerrard

The Latrobe Valley sits on the lands of the Brayakaulung clan of the Gunaikurnai nation. It comprises three major towns—Moe, Morwell and Traralgon—clustered in the central west of the Gippsland region in east Victoria. Fertile rolling hills, thick temperate forests, high rainfall and large reserves of brown coal have made it a heartland for industries including logging, dairy farming and coal-fired power production.

The development of 'The Valley' from the early 20th century, 'arose from the desire of the Victorian government to exploit its brown coal resources for the production of electrical power' (Tomaney and Somerville, 2010: 34). Cheap power produced in The Valley fuelled Melbourne and Victoria's industrialisation (Gibson, 2001), a process overseen by the State Electricity Commission of Victoria (SECV), a publicly owned body responsible for developing Victoria's entire energy system.

The brown coal mined in The Valley is an inefficient and highly polluting fuel that is also unsuitable for export, making the Valley's power industry a major source of Victoria's carbon emissions. Today, the Valley's three coal-fired power stations, each fed constantly from massive co-located open cut coal mines, provide 70% of the state's electricity (Kolovs, 2022). Yet the stability of coal-fired power is faltering and a transition towards new energy and economies is now underway.

Transition is a contested and complex term. When we talk about transition in this chapter, we see it not solely as a process of energy system change or one overseen by Government policy. Rather, it is a composite of the endeavours of many to build new futures, identities,

DOI: 10.4324/9781003585343-3

stories and hopes for a region. In this way, transition is best understood as a process of ongoing negotiation, enmeshed with notions of place, community, and regional futures (Gerrard, 2024).

How the transition process began

The Latrobe Valley has been a place of transition and transformation ever since the violent dispossession of the Gunaikurnai people from the early 1800s (Alexandra, 2017). Increasingly since the early 2000s, the idea of low-carbon transition has loomed large in the Valley. Decades of government and industry inaction in planning for a future beyond coal led diverse stakeholders and initiatives to pursue a more proactive approach to transition, with many demanding a 'just transition' that would ensure workers and communities are supported through change (Edwards et al., 2022). In this chapter, we identify three key 'events' as catalytic forces in the Valley's unfolding transition.

Privatisation

Under the helm of the SECV, the Valley enjoyed sustained periods of growth, near full employment and prosperity (Wiseman et al., 2017). By the mid-1970s, the SECV provided employment to over 9000 residents (Duffy and Whyte, 2017). But in the 1980s, this apparent stability was disrupted by reforms made across the power industry purported to increase productivity. The reforms made the assets more valuable for future sale (Wiseman et al., 2017), and they were privatised by Liberal Premier Jeff Kennett during the 1990s. By 2002, thousands of workers had lost their jobs and only 1800 remained as direct employees of the power stations and mines (Tomaney and Somerville, 2010). The rapid rise in unemployment and significant depopulation which accompanied it transformed the region from prosperity to poverty, and it became one of Australia's most disadvantaged regions (Duffy and Whyte, 2017).

The enduring scars of power industry privatisation are key to understanding transition in the Latrobe Valley. It 'ruptured the contract' between people and government (Gibson, 2001) and ultimately made talk of transition from coal more difficult (Musil, 2013; Duffy and Whyte, 2017). Over a decade later, another event would have a similarly important impact on the Valley.

The Mine Fire

On 9 February 2014, the Hazelwood coal mine caught fire (Hazelwood Mine Fire Inquiry, 2014). The fire raged for 45 consecutive days and shrouded the Valley, particularly the town of Morwell, which was directly adjacent to the mine, in plumes of toxic smoke. More than just a catastrophic fire, it was Victoria's worst environmental air pollution event on record (Environment Victoria, 2016).

The community rallied in response to perceived neglect from the government and the power station and mine owner and operator GDF Suez (later rebranded as Engie). An initial meeting called 'Disaster in the Valley' was attended by 1000 people and a 'petition with over 21,000 signatures' was delivered to the government demanding an 'investigation into the health impacts' (Duffy and Whyte, 2017: 429). From this initial meeting, a community group called 'Voices of the Valley' (VOTV) formed. Early VOTV and community campaigning focused on health impacts and reparations from the mine fire and was critical in forcing two iterations of a State Government Mine Fire Inquiry and a series of legal cases against the mine operators (Preiss and Moore, 2020). VOTV also pointed to the long-term health impacts of economic insecurity and despair and became a vocal advocate for a 'just transition' for the region (Wattchow, 2016).

The closures

Energy Brix

Morwell Power Station and its associated coal briquette factory ('Energy Brix') closed in August 2014, just 6 months after the fire broke out. Despite a bailout package of $50 million from the Federal government, 70 staff lost their jobs when the station closed (Whitson, 2014). Blamed on falling wholesale electricity prices and dwindling demand for briquettes (Whitson, 2014), this closure was the first in the region without replacement coal infrastructure on the horizon.

Hazelwood departs

Less than three years after the mine fire, Hazelwood power station, closed in April 2017 with only five months' notice. Australia's most polluting power station, with eight chimney stacks visible from the main street of Morwell, Hazelwood's closure meant its 450 direct employees faced immediate job losses, along with some 300 contractors who depended on it (Wiseman et al., 2017). If the mine fire made 'transition' less of a taboo term, Hazelwood's closure mainstreamed it, albeit with some resistance (see Gerrard, 2024). Conversations and policy decisions that had been too hard suddenly became necessary. Transition became real.

Key actors

A wide variety of actors have initiated, supported, opposed and resisted transition in the Latrobe Valley.

Federal government

As discussed in Chapter 1, Australian federal governments have until very recently contributed very little towards a coordinated low-carbon transition. Climate change and energy policy under conservative governments since 2014 has been 'marked by stasis, uncertainty and conflict' (Wiseman et al., 2017: 6), with leaders instead acting as protagonists in fomenting a culture war around such policy (Brett, 2020). This context and uncertainty have inhibited place-based planning and exacerbated local fears about transition.

More recently, the Albanese government has established a Net Zero Transition Authority to assist national transition planning and coordination. It has also announced Australia's first offshore wind zone off the coast of Gippsland, providing alternative energy jobs in the region.

State government

Following the announcement of Hazelwood's closure, the Andrews Labor Government played a lead role in supporting the Valley's transition. It immediately mobilised an integrated $266 million package of labour market, infrastructure investment and economic renewal policies and programmes (Perkins, 2021). Key to the ongoing delivery of these programmes was the establishment of the Latrobe Valley Authority (LVA), a decentralised arm of state government. This package has been considered broadly successful in ameliorating the impact of Hazelwood's closure and assisting longer-term economic development (Wiseman et al., 2020).

State government energy policy continues to guide on-the-ground transition agendas. Recent developments include the revitalisation of the State Electricity Commission and the declaration of Gippsland as one of six Victorian renewable energy zones. The government has also introduced laws requiring a minimum 5-year notice period for future power station closures, and the State Government is responsible for authorising mine rehabilitation plans, which are critical to long-term community and environmental health outcomes as well as future economic opportunities.

Local government—Latrobe City Council

The Latrobe City council, like many local government authorities in fossil fuel producing regions, has been and remains in a constrained position in managing transition because most policies affecting the region's mining and energy industry do not fall within council jurisdiction (Weller, 2012). Many elected voices on Council have also been vocal supporters of the coal industry, an attachment which 'stems from its link to economic prosperity and security' (Gerrard, 2024: 147). Despite this, the council has been relatively proactive in discussions about low-carbon transition, albeit often effectively advocating for 'change without change' (Musil, 2013) by assuming some ongoing role for coal.

Council engagement and community consultation around 'transition' accelerated after Hazelwood's closure announcement, including the publication of *A Strength Led Transition* (Latrobe City Council, 2016). Alongside advocating for hospital, rail and other infrastructure upgrades, this document still posited coal-based products as a feature in the region's economic future (Latrobe City Council, 2016). Subsequent reports and policies have outlined Latrobe City Council's transition priorities.[1]

Community: local organisations, grassroots participation and absentees

Community organisations play an important role in shaping the region's transition and the LVA has actively sought community input to inform its work. Community members have also played an important role in shaping the transition discussion by changing local narratives of a future beyond coal (Gerrard, 2024).

First Nations organisations like GLaWAC (Gunaikurnai Land and Waters Aboriginal Corporation) are fundamentally concerned about regional economic and environmental change but are rarely recognised in transition proposals. GLaWAC is developing its own Renewable Energy Strategy, noting that the energy sector in Gippsland has 'taken irreplaceable cultural heritage from us and impacted our connection to Country' while very little has been received 'in return to remedy' these impacts (GLaWAC, 2023).

However, significant sections of the Latrobe Valley community are *not* often engaged or included in transition narratives or activities, or face barriers to participating in transition activities and processes (Gerrard, 2024). Despite residing in a 'coal region', many have no direct relationship to the power industry, nor have they directly benefitted from transition initiatives. These include the Valley's large multicultural and faith communities as well as residents who experience complex disadvantage.

Unions

With a historical reputation for industrial conflict (Tomaney and Somerville, 2010), the Valley hosts headquarters of the state's mining and energy union (CFMEU Mining & Energy Division) along with branches of the Australian Manufacturing Workers Union (AMWU) and Electrical Trades Union (ETU).

As democratic worker-oriented organisations holding differing views within their membership, local union branches have engaged with 'transition' in various ways. The Gippsland Trades and Labour Council (GTLC) has often been proactive in supporting economic diversification and sustainable alternative jobs to ensure 'we get the new industries in place before the chains go on the gates of the old one' (GTLC President John Parker, in Chubb, 2014: 30). GTLC pursued a 'Just Transition Strategy' from 2005 onwards and contributed to various reports and transition committees (Fairbrother et al., 2012; ACTU (Australian Council of Trade Unions), 2016). Such efforts helped lay the groundwork for the post-Hazelwood Worker Transition Centre and Worker Transfer Scheme (Snell, 2018).

Environmental non-government organisations (ENGOs)

Often associated with campaigns to 'shut down' power stations, many environmental organisations have also been active voices in calling for a 'just transition' for workers and community members (Martinelli et al., 2016). In recent years, Environment Victoria, Environmental Justice Australia, Friends of the Earth Australia, the Australian Youth Climate Coalition (AYCC) and 350.org have all advocated for renewable energy developments and just transition policies and encouraged community engagement with mine rehabilitation planning.

Education providers

Research institutions have long taken an interest in the Valley, assisting with regional skill audits, worker redeployment assessments, and (re) training programmes (e.g. Fairbrother et al., 2012; TAFE Gippsland, 2023). RMIT and Melbourne University also co-led the LVA Smart Specialisation regional development programme.

Industry

Industry players both within and outside the electricity sector have a complex relationship with the Valley's transition, acting as both brakes on change and catalysts for change.

On one hand, the privatised fossil fuel industry has played a significant role in blocking climate and transition policies by lobbying governments and by stoking 'jobs vs environment' narratives through 'job blackmail' threats of plant closures (Chubb, 2014; see Evans and Phelan, 2016; Wilkinson, 2020). The owners of Hazelwood and Yallourn power stations staunchly opposed carbon pricing policies in the 2000s–2010s, yet also made offers to close *if the government* paid them enough to do so (Snell and Schmitt, 2012). While such 'purchased closure' proposals were never implemented, the industry still extracted enormous cash commitments from federal governments as compensation for proposed and enacted carbon policies (Chubb, 2014; Wiseman et al., 2017).

On the other hand, industry has been a lead instigator of transition in the Valley, most markedly in Engie's decision to rapidly close Hazelwood (Mercier, 2020; Reeves et al., 2022). At closure, the cost of essential safety repairs at the plant was estimated to be $400 million (Lazzaro, 2016) and Engie could attain a higher sale price for its Loy Yang B power station without the competing electricity supply from Hazelwood in the national energy market. Engie also recently commissioned a large battery at the Hazelwood power station site. Likewise, Yallourn Power Station owner EnergyAustralia has announced an earlier closure and plans to build a utility-scale battery in the Valley (Whittaker, 2021). AGL has since made similar announcements regarding its Loy Yang A Power Station.

Other industries in the region are also undergoing transition, contributing to persistent local 'jobs vs environment' narratives. The Maryvale Paper Mill in Morwell has a larger workforce than Hazelwood. It has sourced native timber from the Gippsland-wide forestry sector, making it a key actor in Australia's long-running 'Forest Wars'. While the mill will continue to operate, the native logging sector is subject to its *own* closure and associated State Government transition assistance package (Schapova and Symons, 2023).

The renewable energy industry is growing its presence in the Valley and across Gippsland. The Valley now hosts several successful solar installation businesses and small-scale renewable energy manufacturers. There is $54 billion worth of large-scale renewable projects currently in development or planning in the wider Gippsland region (Newman, 2023).

Key decision points and actions

Privatisation—old capabilities gone, new incentives, players and barriers

The radical 1990s restructuring and privatisation of Victoria's power industry—itself a traumatic 'transition'—shaped the approach to and perception of future transitions in three key ways.

Firstly, it changed the logics by which the industry operated. Rather than a coordinated public service, the energy industry is run by private companies in ways that maximise profit, rather than prioritise social or environmental needs. As manager of the state's energy system, the SECV had been required to plan for a reduction of greenhouse gas emissions. As a result, by the 1990s it had identified initiatives including demand management, energy efficiency and alternative technology development (Christoff and Low, 1999). This long-term sector-wide energy planning and management was abandoned after privatisation. Power stations became 'profit centres' that owners defended to ensure financial returns, and thus became institutional barriers to adopting alternative technologies (Snell and Schmitt, 2012).

Secondly, privatisation shattered the community's trust in government. This led to deep scepticism towards subsequent government and industry plans and promises (Wiseman et al., 2020). The ensuing 'social and economic disadvantage' of the Valley is 'connected to an ongoing narrative of despair rooted in this perceived mistreatment from government, industry and authority' (Duffy and Whyte, 2017). This persistent narrative shapes how prepared people are to envisage a future beyond coal, particularly if it is initiated by anyone 'external' to the Valley (Musil, 2013).

Thirdly, the transfer of ownership of Victoria's energy infrastructure to multinational corporations had the effect of 'internationalising' significant parts of the Latrobe Valley's economic base and its management (Snell and Schmitt, 2012). Power stations changed from local, service-focused operations to relatively small parts of global profit-oriented portfolios. In this context, with less than 1000 workers, Hazelwood was a relatively small and 'expendable' asset that could be threatened with closure in a 'game of brinkmanship' against government climate policy (Chubb, 2014). Privatisation was the basis on which Hazelwood's eventual closure was decided in Engie's boardroom in France.

Demonising Hazelwood and shouting for the shut

From the mid-2000s many in the climate movement directed efforts towards closing the Valley's high-polluting coal industry. Hazelwood was a particular focus, since it was 'Australia's dirtiest', responsible at the time for 15% of Victoria's total greenhouse gas emissions (Fyfe, 2005). Sustained campaigning saw Hazelwood become a powerful symbol and key battlefield of Australia's climate wars (see Chapter 1). Campaigns to 'Switch off Hazelwood' rose to prominence in the late 2000s, and two mass protests attended by hundreds of people took place at the power station's gates in 2009 and 2010.

Images of predominantly city-based activists descending on a regional workplace typically highlighted simplistic narratives of conflict between 'environmentalists' and 'workers', between city and region. This reinforced long-held 'jobs vs environment' narratives, and as a result the 'Switch off Hazelwood' campaign struggled to positively engage Latrobe Valley participants. When in 2010 the campaign shifted its message by rebranding to 'Replace Hazelwood', its by-lines of 'Switch on Renewables' and 'just transition' were either drowned out or perceived as tokenistic.

Ultimately, the campaign was instrumental in constructing the Valley's coal industry as 'dirty' (Tomaney and Somerville, 2010). It contributed to various state and federal policies to phase out parts of the Valley's coal power generation and helped cast Hazelwood as a global climate pariah. However, the spectacle also contributed to negative local attitudes towards climate change policy, 'green' groups and 'transition'.

Advocacy and articulating vision and hope

There have been countless instances where people and organisations have decided to instigate and persist with transition planning and preparation efforts, often swimming against the tide and in hostile waters. Positive visions and tangible projects have helped to cut through the fear and cynicism surrounding transition that looms like the Valley's famous winter fog.

Local environmental organisations like the small but persistent Latrobe Valley Sustainability Group (LVSG) and nearby Baw Baw Sustainability Network (BBSN) have long called for climate and transition policies. The influential Gippsland Climate Change Network (GCCN) has engaged constructively with local government and business, developed extensive energy education programmes, and co-delivered the State-funded Latrobe Valley Community Power Hub. GCCN's 2022 and 2023 'Gippsland New Energy Conferences' sold out. State and national environmental NGOs have also organised multiple 'listening tours' (ASEN, 2014), public forums (Latrobe Valley Express, 2015; French, 2016); 'just transition' research reports (Martinelli et al., 2016); high school 'Climate Justice Summits'; documentary films (such as Power On, 2023) and the 'TRANSFORM (renewable jobs) Expo'.

Established to unite union and green groups in 'breaking down the jobs vs environment debate', Earthworker has worked to create more sustainable and democratic economies. Together with union, academic and industry figures, Earthworker prepared a 'Solar, Wind & Water Industry Plan' in 2000. Victorian AMWU and ETU branches co-launched the plan, and subsequently supported a wind turbine manufacturing operation in Morwell. The local CFMEU branch and GTLC later assisted Earthworker's business plan for solar hot water manufacturing, a cooperatively owned enterprise that came to fruition in Morwell a decade later (Morrison, 2021).

Other initiatives to create positive regional visions include the VOTV's grassroots transition plan for 'jobs and hope'[2]; Communities Leading Change (2021) transition conversations and magazine; ReActivate Latrobe Valley's (2018) grounded provocations to re-imagine the Valley's towns, landscapes and communities (Gray, 2015; Great Latrobe Park, 2023); Earthworker's 'Walk with the Valley'; the 'ever optimistic' local storytelling of Gippslandia (2023); art projects like Coal Hole (PollannaR, 2020); films such as Our Power (2019); and the ongoing Coal Face Podcast (Life After Coal, 2023) and work by Community Power Agency and The Next Economy. Efforts like these have helped transform conversations about transition from uncertainty and fear to hopeful possibility.

The endeavours described above likely had increased political traction due to another community effort: the campaign for independent Morwell candidate Tracie Lund in the 2014 State Election (Gray, 2014). The grassroots campaign in the wake of the Hazelwood Mine Fire took the previously safe National Party seat and made it one of the most marginal in the state (Latrobe Valley Express, 2014).

Hazelwood's closure and the response

Despite the many efforts described above to get transition on the agenda, transition was no longer debatable, but present and inevitable. Just before the announcement in mid-2016 many in the Valley still saw no need for or benefit in transition. The region still provided most of Victoria's electricity, industry regularly reassured that no closures were imminent and that there was 'hundreds of years' worth of coal' still in the ground (Hughes, 2018).

The State Government's response to Hazelwood's closure was notable for its speed, scale, and long-term capacity-building focus. The establishment of the LVA was a central pillar. Importantly, the LVA's mandate was (and is) to tackle *both* immediate transition needs—such as worker re-training and deployment, and business supply chain adjustment—*and* the longer-term challenge of regional diversification and revitalisation. Drawing on the European regional development strategy of 'Smart Specialisation', the LVA has facilitated collaboration across four key regional themes: Food and Fibre, New Energy, the Visitor Economy, and Health and Wellbeing (Goedegebuure et al., 2020). Unlike many bureaucratic bodies it was initially empowered with a degree of autonomy to administer its programmes and has been staffed by many local people with relevant experience.

Short- and medium-term initiatives included the creation of a Worker Transition Centre providing one-on-one support related to skills, training, financial advice, and re-employment assistance. A Worker Transfer Scheme, long proposed by unions, was also established to facilitate early retirements for older workers in the Valley's remaining power stations. Various business support and grant programmes were provided to minimise the impact on local businesses and to incentivise local employment generation (Wiseman et al., 2020).

Community infrastructure and investment funds were provided for new and upgraded education facilities, a new Morwell 'Hi-Tech Precinct', public and social housing upgrades, community sporting infrastructure and the construction of the Morwell 'Govhub' hosting 300 public service jobs (Mercier, 2020). This well-resourced, well-integrated package of labour market, infrastructure investment and economic renewal policies has been broadly considered successful in softening the impact of Hazelwood's closure while assisting longer-term economic development (Wiseman et al., 2020). Briggs and Mey (2020: 43) calculated that the LVA has helped create 'more than 2,500 new jobs and helped generate more than A$99 million of private investment in the Latrobe Valley', with the overall unemployment rate in the region dropping by 3.7% between September 2016 and 2020.

Three remaining power stations mean long-term transition efforts are essential. Yet in May 2024, the Victorian state government announced that the LVA would be subsumed into Regional Development Victoria by the end of the year. It is unclear which of its functions, plans and operations will live on in this new form, but hopefully the LVA's strengths of place-based capacity and network building will be prioritised.

Key opportunities missed

The story of the transition in the Latrobe Valley prior to Hazelwood's closure was characterised as much by *inaction* as by action. In keeping with the lack of proactive policy or public intervention that been typical of neoliberal governments for decades, transition plans were either absent or abandoned or consisted merely of 'business as usual'. As a result, the Latrobe's experience has been one of countless missed opportunities.

Missed deadlines

After building the power station in the 1960s, the SECV announced in 1992 that Hazelwood would be retired in 2005, as part of orderly infrastructure retirements and replacements. After Hazelwood's privatisation though, its new owners had an interest in keeping the plant running to ensure a return on their investment. In the absence of the SECV's state-wide energy planning, the 'free market' for electricity saw insufficient new generation capacity built to replace Hazelwood on schedule.

Instead of using 2005 as the deadline around which to plan replacements, as the year approached the State Government allowed an extension of Hazelwood's operations by exempting greenhouse gas emissions from consideration. A 2004/2005 state planning panel assessed that 'given the lead-time for alternative technologies, the absence of significant [electricity] demand management', the proposed extension to the operating life of the mine and power station was 'the most economical alternative for the supply of base load electricity to Victoria and the National Electricity Market' (Planning Panels Victoria, 2005).

Staged closures abandoned

There were also two key moments in the decade before Hazelwood's rushed closure when a more orderly retirement might have been realised. Both were abandoned in the heat of Australia's 'Climate Wars'. Firstly, in the lead up to the 2010 State election, Premier John Brumby announced a plan to 'purchase' a gradual staged closure of Hazelwood (Lesman et al., 2011). However, the Liberal-National Party opposition ran a strident and effective campaign against the plan, playing on fears of local job losses (Wiseman et al., 2017). Upon entering government, they ditched the plan entirely.

Secondly, development of the Gillard Federal Labor government's 'Clean Energy Future' (or 'Carbon Tax') package in 2010/2011 included a 'contract for closure' programme, supported by some power station owners, that would have paid coal-fired power stations to close in coordinated stages. A proposed $200 million 'regional structural adjustment fund' would also have helped affected communities like those in the Valley to diversify and adjust (Chubb, 2014). Unfortunately, both proposals were dropped in 2012 when flaws in setting the carbon price, plus handsome corporate compensation, 'breathed new life' (and financial value) into coal-fired generators (Chubb, 2014: 236–237).

The 2007 election of the Rudd Labor Government, and its assurances of regional structural adjustment support, heralded a time when public support for climate action was at an all-time high. The promise of carefully phased closures and transition assistance helped allay some fears in the Valley, but the ultimate 'failure of power station closure plans had a dramatic effect on negotiations for a smooth transition to a low-carbon future in the Latrobe Valley' (Chubb, 2014: 237).

Choosing business as usual

The lack of government planning for an orderly transition from coal has amounted to a de-facto promotion of (coal) business as usual. In recent decades, millions of dollars have been promised, spent, and ultimately squandered by state and federal governments on failed 'clean coal' projects in Gippsland. These include coal-to-oil or coal-drying-for-export processes (Millar and Schneiders, 2018), carbon capture and storage (Arup, 2009) and more recently coal to hydrogen (Lazzaro, 2018)—projects which have already failed, or the viability of which are highly questionable (Readfearn, 2021).

The same can be said for enormous cash handouts and tax breaks to the Valley's coal generators as 'compensation' for carbon pricing—amounts dwarfing any support offered to affected workers and communities (Chubb, 2014; Wiseman et al., 2017). In 2012, Engie paid itself a $1 billion dividend while simultaneously complaining that the $500 million in cash and tax credits it had just received from the Gillard Government was not enough compensation for the carbon tax (Chenoweth, 2017). Hazelwood workers rightly raised concerns that these public funds—given without any accompanying conditions—were sent straight overseas (Chubb, 2014: 238–239).

In contrast, countless alternative choices have been available to forward-thinking administrations. In the late 1990s, ten wind turbine generators were manufactured in Morwell for export to Japan. This initiative of local start-up REAP-Primergy with support from Earthworker and unions was noticed by the International Energy Agency, which observed that government policy 'could see this manufacturing capability maintained and possibly expand' (IEA, 2001). However, the Howard Federal Government was opposed to industry policy and renewable energy and the Bracks Victorian State

Government too hesitant. When private capital failed in 2001, the project ended.

A decade later, the State Government 'pledged its support' to Earthworker's Morwell solar hot water manufacturing proposal (Lord, 2010) before a change of government and bureaucratic cold feet dashed hopes again. While this project later commenced through grassroots investment and persistence (Morrison, 2021), earlier government assistance for these and other initiatives may well have better prepared the Valley for its now unfolding transition.

Key lessons from the Latrobe Valley's transition experience

Transition is contextual, shaped by people, place, history and culture

The story of transition in the Valley shows that it is not a 'one off event'. Rather, it is a process embedded in local historical, political and cultural contexts (Duffy and Whyte, 2017). Many locals see the current discussions of transition as a 'continuation of a longer tradition of major setbacks to the regional economy' (Holm and Eklund, 2018: 71). This sense of a protracted transition is most keenly observed in how trauma of industry privatisation still ripples through the Valley (Reeves et al., 2022). There is also a powerful legacy of communities in the Valley feeling as though things get done to them, rather than with them (Tomaney and Somerville, 2010; Duffy and Whyte, 2017):

> While some elements of the Valley's historical context—including a strong tradition of worker voice and union leadership—helped to enable the transition response, other elements acted to inhibit change. 'Jobs vs environment' narratives associated with local logging and coal industries remain prevalent, while the legacy of privatisation and enduring climate policy uncertainty often still make it difficult to even discuss transition.
>
> (Musil, 2013)

Cultural and regional identity also play a key role in how a region responds to change. There is a dominant image of the Latrobe Valley as an industrial coal and energy region with the male blue collar coal worker at its core (Tomaney and Somerville, 2010; Farhall, 2021). While coal and coal workers have played an integral role in the prosperity of the region, shifting this image and allowing the formation of new regional identities is an important part of transition. Furthermore, this ingrained cultural and regional identity tends to render feminised industries like health, aged-care and community services as subordinate, despite these sectors now being the biggest employers in the region. This makes transition a highly gendered issue (Farhall, 2021). The ongoing pursuit of a brown coal hydrogen industry in the Valley links to a wider trend in Australia that prioritises extractive industries and inflates their economic contribution (Brett, 2020).

Similarly, the Valley's identity as an 'industrial region' carries a predilection to view economic futures through a familiar lens. Many in the Valley seek another major industry to arrive and 'save' the region by replacing one industrial employer for another near identical one. This regional development strategy of 'chasing smokestacks' to 'attract the golden egg of corporate employment' is rarely successful (Cameron and Gibson, 2005). The State Government's recently failed efforts to lure an electric vehicle manufacturer to Morwell (Eddie and Preiss, 2021) and the retreat of the new Morwell Hi-Tech precinct's multinational anchor tenant (Whittaker, 2019) demonstrate an ongoing attachment to this often-problematic strategy.

Local voices, leadership and community ownership are vital

Given long-held local cynicism and perceptions of power being imposed on the region by 'outsiders' (be they governments, businesses, activists or others), genuine local participation in transition policies, programmes and responses is essential. As we have outlined, local voices need to be diverse, and efforts need to be made to seek out underrepresented groups. Regional transition affects everyone and is therefore best delivered in inclusive and fair ways.

Local leadership—working in formal and informal spaces—has been a pivotal feature in the Valley's transition to date. From the early leadership of the GTLC on regional diversification and just transition strategies (Chubb, 2014) to the community organising and education of VOTV and GCCN and the economic experimentation of Earthworker, there is a long tradition of local groups and individuals facilitating challenging conversations and leading change (Gerrard, 2024).

Scale, diversification and experimentation are key

Defining the scale of transition, particularly a 'just transition', is a persistent problem in theory and practice (Musil, 2013; Stevis, 2021). Is the transition solely to replace the Valley's dwindling brown coal industry, or should it include the wider Gippsland region as native timber and oil and gas industries also decline, and entrenched socio-economic disadvantage persists?

The Valley's power industry has long been at the centre of Victoria's electricity grid, with the provision of baseload power part of the region's industrial identity. However, a future renewable-powered electricity grid will be less centralised, more dependent on dispatchable than baseload power, and less labour intensive (McBain, 2016). Those who approach the Valley's challenge as primarily an *energy transition* are therefore commonly stuck seeking an impossibly neat swap from local coal energy jobs to equal quantities of local renewable energy jobs (Musil, 2013). Electricity generation is certainly part of the Valley's transition, but transition in the Valley is necessarily a question of *diversification*, both of economic activities *and* regional identities.

As discussed above, there remains a rarely fruitful tendency to seek large-scale industries to replace similar ailing ones. Transition approaches that seek only like-for-like replacement jobs for current workers reflect a capitalo-centric way of thinking (Gibson-Graham, 2006) that obscures *already existing* economic diversity and opportunities while neglecting the varied ways other stakeholders can contribute to regional futures. Regional transition is best supported through economic diversification (Cahill, 2022) when all scales and types of enterprise are invigorated and supported.

Long-term planning, coordination and resourcing are essential

The resounding message from Hazelwood's closure is that long-term planning and coordination of transition is essential (Wiseman et al., 2020). With further power station closures scheduled, the region's future depends on ensuring that there is ongoing, well-resourced planning and coordination. While the LVA formed quickly and played a critical role in dealing with immediate repercussions of Hazelwood's closure including seeking local expertise, prior proactive planning with meaningful industry and community buy-in would have negated the need for reactive measures. Ensuring secure ongoing funding for initiatives like the LVA and other long-term diversification strategies is critical. Yet as mentioned earlier, with the LVA's announced absorption into Regional Development Victoria, uncertainty is again on the horizon.

Finally, engagement in transition—visioning, planning, critique, execution—doesn't always have to be in the language of climate, energy or even 'transition'. Given the baggage often carried by the term transition, there are times and places where 'talking about transition' can inhibit engagement, action or change. Many of the initiatives outlined above were significant transition initiatives even (or especially) if not identified as such. They include initiatives that have invited communities to encounter their neighbourhood in new ways, encouraged people to imagine what rehabilitated minescapes or healthier towns could look like, facilitated new connections between organisations and individuals, enrolled participation through art, film, sound or local food, and simply demonstrated that things can be done differently. As the mine fire and Hazelwood's sudden closure showed, disruptions can be tragic and painful, but also fruitful. Being invited (or sometimes prodded) to experience, witness, and attempt different ways of doing and being can precipitate small shifts that can culminate in big transformations.

Notes

1 See www.latrobe.vic.gov.au/Council/Media_and_Publications/Major_Council_Publications
2 www.votv.org.au/transition

References

ACTU (Australian Council of Trade Unions) (2016) *Sharing the challenges and opportunities of a clean energy economy: A Just Transition for coal-fired electricity sector workers and communities*. www.actu.org.au/media/1032953/actu-policy-discussion-paper-a-just-transition-for-coal-fired-electricity-sector-workers-and-communities.pdf (accessed 27 September 2024).

Alexandra J (2017) Water and coal - transforming and redefining "natural" resources in Australia's Latrobe region. *Australasian Journal of Regional Studies* 23(3): 358–381.

ASEN (2014) Mining the truth roadtrips. https://asen.org.au/events/miningthetruthroadtrips (accessed July 2023).

Arup T (2009) Carbon storage identified. *The Age*, 9 December 2009, p. 4.

Brett J (2020) *The Coal Curse: Resources, climate and Australia's future. Quarterly essay*. Australia: Black Inc.

Briggs C and Mey F (2020) *Just transition: Implications for the corporate sector and financial institutions in Australia*. ISF for Global Compact Network Australia and National Australia Bank. Sydney.

Cahill A (2022) *What regions need on the path to net zero emissions*. Brisbane: The Next Economy.

Cameron J and Gibson K (2005) Alternative pathways to community and economic development: The Latrobe Valley Community Partnering Project. *Geographical Research* 43(3): 274–285.

Chenoweth N (2017) Paradise papers: Loy Yang paid $1b dividend to Engie ahead of carbon tax. *Australian Financial Review*, 9 November 2017.

Christoff P and Low N (1999) Recent Australian urban policy and the environment: Green or mean? In I Elander, B Gleeson, R Lidskog and N Low, eds. *Consuming Cities*. London and New York: Routledge.

Chubb P (2014) *Power failure: The inside story of climate politics under Rudd and Gillard*. Melbourne: Black Inc.

Communities Leading Change (2021) *Transitions - stories of Gippsland communities leading change*. www.communitiesleadingchange.org.au/resources (accessed July 2021).

Duffy M and Whyte S (2017) The Latrobe valley: The politics of loss and hope in a region of transition. *Australasian Journal of Regional Studies* 23(3): 421–446.

Eddie R and Preiss B (2021) Electric vehicle deal to create regional jobs collapses. *The Age*, 9 November 2021.

Edwards GAS, Hanmer C, Park S, MacNeil R, Bojovic M, Kucic-Riker J, Musil D and Viney G (2022) *Towards a just transition from coal in Australia?* London: The British Academy. https://doi.org/10.5871/just-transitions-a-p/G-E.

Environment Victoria (2016) *Looking back on Hazelwood*. https://environmentvictoria.org.au/2016/11/16/looking-back-6-moments-campaign-phase-australias-dirtiest-power-station/ (accessed 27 September 2024)

Evans G and Phelan L (2016) Transition to a post-carbon society: Linking environmental justice and just transition discourses. *Energy Policy* 99: 329–339. https://doi.org/10.1016/j.enpol.2016.05.003

Fairbrother P, Snell D, Bamberry L, Condon L, McKenry S, Winfree T, Stroud D and Blake J (2012) *Jobs and skills transition for the Latrobe Valley: Phase 1: Benchmark occupations and skill sets*. Melbourne: RMIT University.

Farhall K, Tyler M and Fairbrother P (2021) Labour and regional transition: Sex-segregation, the absence of gender and the valorisation of masculinised employment in Gippsland, Australia, *Gender, Place & Culture* 28(12): 1755–1777. DOI: 10.1080/0966369X.2020.1858031

Fyfe M (2005) Hazelwood wins worst polluter dishonour. *The Age*, 14 March 2005, Melbourne.

Gerrard E (2024) *Naming, "resisting" and "making": Understanding community experiences of a (just) transition in regional Australia.* [Doctoral Thesis, University of Tasmania].

Gibson K (2001) Regional subjection and becoming. *Environment and Planning D: Society and Space* 19(6): 639–667.

Gibson-Graham JK (2006) *A postcapitalist politics*. Minneapolis: University of Minnesota Press.

Gippsland Climate Change Network (2023) *GCCN Initiatives*. www.gccn.org.au/initiatives (accessed July 2023).

Gippslandia (2023) *About*. https://gippslandia.com.au/about (accessed 30 July 2023).

Goedegebuure L, Wilson B, Coenen L, Schoen M, Fastenrath S, Ward C and Shortis E (2020). *Developing and implementing a smart specialisation approach for Gippsland, Victoria (2018-2020)*. Melbourne Sustainable Society Institute, University of Melbourne, RMIT University's European Union Centre of Excellence for Smart Specialisation and Regional Policy.

Gray D (2014) Battle for Morwell heats up as independent Tracie Lund launches campaign. *The Age*, 7 October 2014.

Gray D (2015) Sunflowers in full bloom in the middle of Morwell in the Latrobe Valley. *The Age*, 26 February 2016.

Great Latrobe Park (2023) *About the Project*. https://glp.org.au/about-the-project/ (accessed July 2023).

Gunaikurnai Land and Waters Aboriginal Corporation (2023) *Renewable energy transition*. https://gunaikurnai.org/our-economy/renewables/ (accessed July 2023).

Hazelwood Mine Fire Inquiry (Victorian Government) (2014) *Hazelwood mine fire inquiry report*. Analysis, Policy & Observatory. https://apo.org.au/node/41121

Holm A and Eklund E (2018) A post-carbon future? Narratives of change and identity in the Latrobe Valley, Australia. *BIOS - Zeitschrift für Biographieforschung, Oral History und Lebensverlaufsanalysen, 31*(2): 67–79. https:// doi.org/10.3224/bios.v31i2.06?>

Hughes A (2018) Australian Resource Reviews: Brown Coal 2017. Canberra: Geoscience Australia. http://dx.doi.org/10.11636/9781925297997 (accessed May 2019).

IEA (International Energy Agency) (2001) Wind Energy Annual Report 2000, National Renewable Energy Laboratory, Colorado. www.nrel.gov/docs/fy01osti/29436.pdf

Kolovs B (2022) Greens propose shutting down all Victorian coal-fired power plants by 2030. *The Guardian*. www.theguardian.com/australia-news/2022/aug/15/greens-propose-shutting-down-all-victorian-coal-fired-power-plants-by-2030

Latrobe City Council (2016) *A strength led transition.* www.latrobe.vic.gov.au/sites/default/files/A_Strength_Led_Transition_-_Latrobe_City_Council_2016.pdf

Latrobe Valley Express (2014) Marginal mission. *Latrobe Valley Express*, 30 November 2014, Morwell.

Latrobe Valley Express (2015) Where to from here? *Latrobe Valley Express*, 25 February 2015, Morwell.

Lazzaro J (2016) Worksafe notices detail extent of repairs needed at Hazelwood. *ABC News*, 1 December 2016. www.abc.net.au/news/2016-12-01/worksafe-notices-detail-extent-of-repairs-needed-at-hazelwood/8082318

Lazzaro J (2018) World-first coal to hydrogen plant trial launched in Victoria. *ABC News*, 12 April 2018. www.abc.net.au/news/2018-04-12/coal-to-hydrogen-trial-for-latrobe-valley/9643570

Lesman B, Macreadie R and Gardiner G (2011) *The 2010 Victorian State Election.* www.parliament.vic.gov.au/publications/research-papers/send/36-research-papers/13686-2010-victorian-state-election

Life After Coal (2023) *Life after coal.* https://lifeaftercoal.com/ (accessed July 2023).

Lord E (2010) Social co-ops to secure future. *Latrobe Valley Express*, 26 July 2010, Morwell.

Martinelli A, Aberle N, Wakeham M, Nadel C and Merory A (2016) *Life after coal: Pathways to a just and sustainable transition for the Latrobe Valley.* Report, Environment Victoria. https://apo.org.au/node/69785 (accessed 27 September 2024).

McBain B (2016) Reliable renewable electricity is possible if we make smart decisions now. *The Conversation*, 1 December 2016. https://theconversation.com/reliable-renewable-electricity-is-possible-if-we-make-smart-decisions-now-68585

Mercier S (2020) *Four case studies on just transition: Lessons for Ireland research series paper no. 15.* National Economic and Social Council. https://ssrn.com/abstract=3694643

Millar & Schneiders (2018) Browned off: $90m "clean coal" program ends as final project collapses. *The Sydney Morning Herald*, 18 April 2018. www.smh.com.au/environment/climate-change/browned-off-90m-clean-coal-program-ends-as-final-project-collapses-20180418-p4zacj.html

Morrison D (2021) Thinking co—operatively. Earthworker Energy's battle for green manufacturing. *Gippslandia,* 4 April 2021.

Musil D (2013) "Greenies vs Coal Communities?" – Geographies of deliberation in Victorian climate change and energy debates. [Undergraduate Honours dissertation, University of Melbourne]. https://minerva-access.unimelb.edu.au/items/e903ed5a-3f7f- 5cd6-ac10-901f204375d1

Newman C (2023) *Gippsland Procurement Power Initiative harnessing the renewable energy transition to transform social and economic outcomes in Gippsland.* ArcBlue. https://lva.vic.gov.au/business-and-worker-support/Gippsland-Procurement-Power-Initiative-Report.pdf

Our Power (2019) Our power: Reconnecting our communities [Film]. https://ourpowerdoco.com/watch/ (accessed July 2023).

Perkins M (2021) Power shift: The Latrobe Valley looks for a new future, again. *The Age,* 27 September 2021. www.theage.com.au/national/victoria/power-shift-the-latrobe-valley-looks-for-a-new-future-again-20210923-p58u7i.html

Planning Panels Victoria (2005) Hazelwood West Field EES, La Trobe Planning Scheme Amendment C32, Final Panel Report 16, 4 March 2005. www.austlii.edu.au/cgi-bin/viewdoc/au/cases/vic/PPV/2005/16.html

PollyannaR (2020) The art of rehabilitation – The Coal Hole. A pile of dirt or a cool arts project? *Gippslandia*, 22 January 2020

Power On (2023) Power on: Gippsland's energy transition [Film]. www.poweronfilm.com.au/about (accessed July 2023).

Preiss B and Moore G (2020) Hazelwood operator fined more than $1.5m for 45-day mine fire. *The Age*, 19 May 2020. www.theage.com.au/national/victoria/hazelwood-operator-fined-more-than-1-5m-for-45-day-mine-fire-20200519-p54ucv.html

Readfearn G (2021) Australia's only working carbon capture and storage project fails to meet target. *The Guardian*, 12 November 2021

Reeves J, Baumgartl T, Morgan D and Green M (2022). *Community capacity to envisage a post-mine future: Rehabilitation options for Latrobe Valley brown coal mines*, pp.173–186. 10.36487/ACG_repo/2215_09.

Schapova N and Symons B (2023) Victoria's native timber logging communities shocked by acceleration of industry's end. *ABC News*, 25 May 2023 www.abc.net.au/news/2023-05-25/death-of-timber-industry-rocks-victorian-logging-communities/102385506

Snell D (2018) 'Just transition'? Conceptual challenges meet stark reality in a 'transitioning' coal region in Australia. *Globalizations* 15(4): 550–564, DOI: 10.1080/14747731.2018.1454679

Snell D, and Schmitt D (2012) 'It's Not Easy Being Green': Electricity corporations and the transition to a low-carbon economy. *Competition and Change* 16(1): 1–19.

Stevis D (2021) The globalization of Just Transition in the world of labour: The politics of scale and scope. In L Mello e Silva, E Estanque and H Augusto Costa, eds., Transnational labour struggles and political repertoires. *Thematic Issue of Tempo Social* 33(2): 57–77.

TAFE Gippsland (2023) TAFE Gipplsand and Fed Uni Sign MOU. www.tafegippsland.edu.au/about/mediacentre/2019/tafe_gippsland_and_fed_uni_sign_mou

Tomaney J and Somerville M (2010) Climate change and regional identity in the Latrobe Valley, Victoria. *Australian Humanities Review* 49: 29–47.

Wattchow K (2016) The future of the Latrobe Valley: Community leadership on Just Transition. *Chain Reaction* 128: 22–23.

Weller S (2012) The regional dimensions of the "Transition to a Low-carbon Economy" the case of Australia's Latrobe Valley. *Regional Studies* 9: 1261.

Whitson R (2014) Coal-fired power station to close in Latrobe Valley, despite $50 million federal bailout. *ABC News*, 29th July 2014. www.abc.net.au/news/2014-07-29/vic-coal-power-station-to-close2c-despite-bail-out/5633580

Whittaker J (2019) Fujitsu pulls out of planned Victorian Government-backed tech precinct. *ABC News*, 8 October 2019. www.abc.net.au/news/2019-10-08/fujitsu-out-latrobe-valley-hi-tech-precinct/11581432

Whittaker J (2021) Hydrogen production from coal generates export hopes, emissions fears. *ABC News*, 12 March 2021. www.abc.net.au/news/2021-03-12/hydrogen-from-coal-production-begins-la-trobe-valley/13241482

Whittaker J and Symons B (2023) The first big battery at an Australian coal site goes live at Hazelwood power station. *ABC News*, 14 June 2023. www.abc.net.au/news/2023-06-14/first-big-battery-at-ex-coal-site-goes-live/102477230

Wilkinson M (2020) *The Carbon Club: How a network of influential climate sceptics, politicians and business leaders fought to control Australia's climate policy*. Sydney: Allen & Unwin.

Wiseman J, Campbell S and Green F (2017) *Prospects for a "just transition" away from coal-fired power generation in Australia: Learning from the closure of the Hazelwood Power Station* (CCEP Working Paper 1708). Crawford School of Public Policy, Centre for Climate Economics and Policy, Australian National University.

Wiseman J, Workman A, Fastenrath S and Jotzo F (2020) *After the Hazelwood coal fired power station closure: Latrobe Valley regional transition policies and outcomes 2017-2020.* (CCEP Working Paper 2010). Crawford School of Public Policy, Centre for Climate Economics and Policy, Australian National University.

4 Centring Country and community in the transition of Collie, Western Australia

Naomi Joy Godden, Georgia Beardman, Mehran Nejati, Jaime Yallup Farrant, Evonne Scott, Leonie Scoffern, James Khan, Joe Northover, Lynette Winmar, Phillip Ugle, Keira Mulholland, Stevie Anderson, Jayla Parkin and Angus Morrison-Saunders

Introduction

The regional town of Collie, population 8812, is located on the *Boodja* (Country or land) of the Wilman (Freshwater) Noongar people in the southwest region of Western Australia (WA). In Wilman *wongi* (language) the location of the townsite is called *Koolinup* which means 'swimming in water'. For millennia, *Beelagu* (the river people) have swum in *Mardalup,* the Collie River.

This area of Wilman *Boodja* was colonised in the 1880s after settlers found coal in the Collie basin. For over 125 years, Collie has been the epicentre of coal extraction and coal-fired energy production in WA, with two coal mines and three coal-fired power stations (Figure 4.1). However, in 2017 the WA Government decided to retire the state-owned Muja and Collie coal-fired power stations by 2030. Two units of Muja (A and B) were retired in 2017, Muja C is slated for closure by 2025 and Muja D by October 2029. Collie power station will retire by October 2027. In 2020, a Just Transition Working Group (JTWG) and the WA Government developed *Collie's Just Transition Plan* to fund and implement a just transition for affected workers and

DOI: 10.4324/9781003585343-4

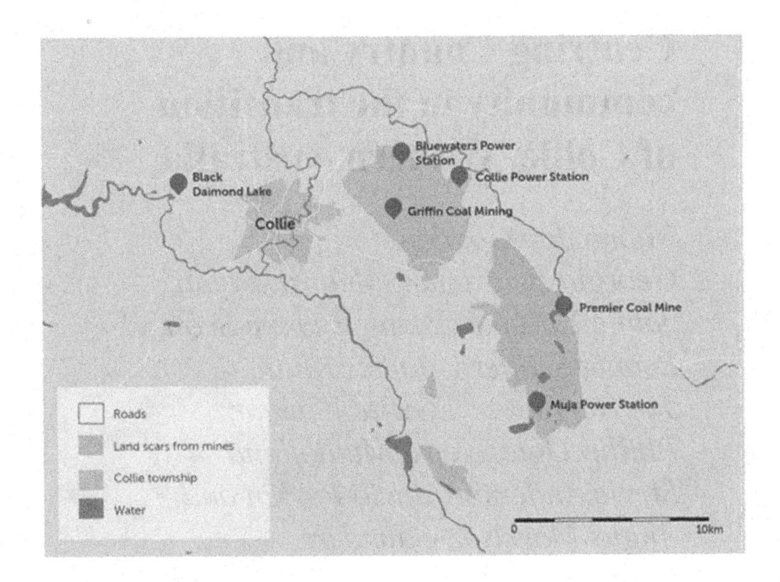

Figure 4.1 Map of Collie and surrounds.

Source: Beyond Zero Emissions (2019)

the wider community away from its economic dependence on coal (Department of Premier and Cabinet [DPC], 2020). To date the WA Government has committed over $660 million to Collie's transition.

This chapter examines the transition process in Collie drawing on a feminist participatory research project led by a collaboration of Wilman Elders, Collie community members, Climate Justice Union WA and Edith Cowan University's Centre for People, Place and Planet. This project started in 2021, building on decades of community action and organising. The project aims to support the Collie community to achieve a transition that centres *Boodja* and First Nations peoples, reduces inequities for current and future generations, and leaves no one behind. Over 200 community members have participated in project activities including four social mapping workshops (N= 40), an online survey (N=148), a data analysis workshop (N=17), a writing workshop (N=10) and an Elder-led meeting On Country with policymakers (N=7).

The phase-out of coal-fired power generation in Collie along with the impacts of climate change will have major ramifications for

Boodja and the people of the town and region. This chapter examines how transition planning has proceeded to date and argues that *Boodja* and community should be the focus of just transition planning and implementation.

Where and when did the transition process begin?

As highlighted by other chapters in this volume, the current transition from coal-fired energy production in Collie is just the latest in a long history of transition affecting *Boodja* and the community of Collie.

The Wilman people of the Bibbulmun Nation have coexisted on Wilman *Boodja* for tens of thousands of years. Their ancient wisdom and law enable them to care for, regenerate and heal the interconnectedness of *Boodja* (Country), *Bilya* (River), *Kep* (Water) and *Moort* (Family or People). The *Ngarngungudditj Waugal* (hairy faced snake) is the creation spirit of *Mardalup* (the Collie River) and the freshwaters of this region. Wilman Elders explain that *Mardalup* is spiritually and culturally significant because 'if we haven't got the water we haven't got a survival kit' (Workshop 3).

Wilman people have overcome many transitions, including a 10,000-year drought during the ice age (Robertson and Barrow, 2020) and massacres by colonisers in neighbouring places such as Pinjarra in the early 1800s. The development of the coal industry in Collie in the 1900s was also a transition. Settlers found coal in 1883 and production commenced in 1898. Wilman land and sites of cultural and spiritual significance were seized by colonists for pastoralism, residential landholdings, and the construction of a train line and town centre. Wilman people were subjected to the *Aborigines Act 1905* which 'laid the basis for the development of repressive and coercive state control' of Aboriginal people (Clark, 2019). This included the forced removal of Aboriginal children from their families, known as the Stolen Generations. Roelands Mission was located on a farm 41 km from Collie, and between 1938 and 1975 it housed over 500 Aboriginal children forcibly taken from their families from all over WA (Find and Connect, 2021).

Soon after coal production began in Collie, the community became economically dependent on the coal market. By 1920, coal mining production had increased to 470,000 tonnes, employing over 800 people (Mining & Energy WA, n.d.). Collie coal was used for WA's energy needs rather than exported, a trend which continues today. The first coal-fired power station was built in Collie in 1931 and Collie's coal plants remain the hub of WA's South West Interconnected System electricity grid. Even the peak coal production period in Collie was characterised by periods of boom and bust that sparked their own transitions, including the closure of Amalgamated Collieries in 1960 and record population growth in the mid-20th century due to post-war power demands.

The present transition process began in earnest in 2017, when the WA Government recognised that the viability of coal-fired power was declining and declared its intention to close the coal-fired power stations the state-owned electricity company Synergy operates in Collie (DPC, 2020). This started a dialogue about transition between industry, community, unions and government, and a JTWG was established in 2019. The JTWG has representatives from industry, unions, and state and local government and the community (represented by the local government). This group developed *Collie's Just Transition Plan* which was launched by the WA Government in December 2020. The plan is described as 'the foundation for what is recognised will be a 10–15 year transition process for Collie' (DPC, 2020).

Who were the key actors?

This section outlines the key actors in Collie's transition planning so far.

WA Government

The WA Government catalysed coal transition in Collie and led transition planning through the Department of Premier and Cabinet (DPC). The WA Government set out a framework to establish the JTWG 'in collaboration with State agencies, local government, industry, worker representative groups and key community stakeholders' (DPC, 2020: 15) and has committed more than $662 million 'to drive economic diversification and create jobs in the Collie region' (DPC, 2022a). This funding is explored further in the next section. DPC also established and hosts the Collie Delivery Unit (CDU), which has staff in Collie and Perth. The CDU 'coordinates, oversees and promotes activities and initiatives that support the future prosperity of Collie' (DPC, 2022a), focusing on economic diversification.

Just Transition Working Group

Collie's transition process has been driven by the JTWG, which comprises representatives from state and local governments, unions and employers. The JTWG is responsible to 'ensure affected people are considered by decision-makers and that early action towards a Just Transition can minimise any negative impacts and maximise positive opportunities' (DPC, 2023a). The JTWG developed a set of principles to guide Collie's transition, which are endorsed by the WA Government (Figure 4.2).

Due to a history of conflict and division relating to Collie's transition from coal, there was some reluctance for the JTWG to be a large coalition. The individual membership of the JTWG is not made public, and some community leaders have expressed difficulties participating in the group due to its processes and communications. As discussed later, our research with the community has found that the formal working group structure for Collie's transition is not yet sufficiently inclusive.

Unions

Union groups have a long history of worker organising and advocacy in Collie, and the Electrical Trade Union, Mining and Energy Union, Australian Manufacturing Workers' Union, and Australian Services Union all have a strong presence, being both represented on the JTWG and shaping transition conversations. For instance, it was lobbying by union groups that led to the inclusion of 'just transition' language in the JTWG and broader transition planning (Tattersall, 2022).

Local government

Local government has been a key actor in the transition process because of its role in administering community infrastructure, programmes and land use planning across the Shire of Collie. It is important to understand, though, that the Shire does not hold decision-making authority regarding Collie's transition, so its influence is limited (Shire of Collie, 2022). The CEO and one councillor are members of the JTWG, representing 'the community'.

Collie's Just Transition Plan

Guiding Principles

1. Encourage sound investments in low emission and job-rich sectors and technologies that attract and maintain local employment opportunities.
2. Recognise, promote and celebrate the history, cultural heritage and invaluable contribution the town of Collie and associated coal and power generation industries have made and continue to make to the State of Western Australia.
3. Respect the rights of those affected by transition to be treated with justice and dignity.
4. Ensure all consultation and negotiations are honest, open and transparent and work towards achieving consensus on goals, timelines and pathways.
5. Recognise that "we're all in this together", and thus share the challenges and opportunities transition brings. The objective is to see that no-one is left behind.
6. Strive to ensure that all those affected by transition are given comprehensive information, opportunity and choice to retrain, reskill/upskill or take an alternative pathway within a reasonable timeframe.
7. Organise local, long-term economic diversification plans that support worthwhile occupations and foster continuous improvement in local living standards for current and future generations.
8. Provide policy, social support and linkages to community and government services for the benefit of all those affected by transition.

Figure 4.2 Guiding principles of Collie's just transition plan.

Source: Reproduced from DPC (2020)

Industry

Major employers are mainly involved in transition planning through membership of the JTWG, but Synergy is more substantially involved as a state-owned energy corporation with local employees and subcontractors for whom it is responsible. Synergy has a transition package for impacted workers including 'skill assessments, personal planning support, redeployment, upskilling, financial planning support, assistance with job search, training programs, funding for approved training programs and preparation for retirement' (DPC, 2020: 18). Other industry actors in the transition include subcontractors and private coal mining companies in the Collie region, some of which have long-standing disputes with local workers and the community. In our project, some community members feel that while Synergy workers are receiving significant support, workers in private coal mines, contractors, and local businesses that provide services to Synergy and other companies are being left behind. Community members also express concern that large companies in Collie are not upholding their responsibilities to support the community through the transition. For example, there are fears that energy and mining companies will withdraw or reduce their sponsorship of local programmes and events, and that new companies involved in lithium extraction may destroy local forests.

Wilman Elders

For more than 65,000 years, Wilman Elders have been caring for Wilman *Boodja, Kep, Bilya* and *Moort,* leading their community and Country through many transitions. In the context of the social, economic and environmental impacts of colonisation in Collie, the transition of Collie's major industry is a key concern for Elders. Wilman Elders have engaged in the transition process by attending various events and participating in some consultation activities when invited. However, the cultural knowledge, priorities and labour of Elders are not (at the time of writing) prioritised in Collie's formal transition processes, and their perspectives are not included in *Collie's Just Transition Plan.*

Community

All members of the diverse Collie community are key stakeholders in the transition. This includes people who are directly affected, such as coal workers, and people who are indirectly affected by the widespread impacts of Collie's economic changes. However, community members recognise that many perspectives and lived experiences of Collie's transition are not centred or prioritised in transition planning, including Wilman Traditional Owners and other Aboriginal people, people on low incomes, people with disability, young people and older people, LGBTQIA+ people, and women and non-binary people.

What were the key decision points and actions?

Collie's current transition launches from a long history of contestation and activism for competing priorities between Traditional Owners, workers, industry and environmentalists. In particular, workers and unions have led various industrial actions to demand fair and decent working conditions for coal workers. In recent years, Indian-owned coal mining company Griffin Coal has been a subject of worker disputes, including a 180-day strike labelled as 'WA's longest running coal dispute' (WA Today, 2018). Union members have also held protests outside the Albermarle lithium processing plant in Kemerton (53 km west of Collie) to secure local jobs and fair working conditions (Jeffers and Cantatore, 2019).

Furthermore, as the main producer of coal and coal-fired energy in WA, Collie has been monitored and targeted by environmental activists. Between the 1990s and 2010s, some environmental demonstrations were held in Collie to protest the impacts of coal-fired energy production on climate change and urge the government to reduce its reliance on coal (see, for example, Towie, 2009). Figure 4.3 highlights some of the key government decisions and actions associated with Collie's transition from coal-fired power so far.

2019: Beyond Zero Emissions report

In November 2019, climate change think tank Beyond Zero Emissions published the report *Collie at the Crossroads: Planning a Future Beyond Coal.* This report was developed in partnership with Wilman Elders and members of the Gnaala Karla Booja Working Party, Collie community

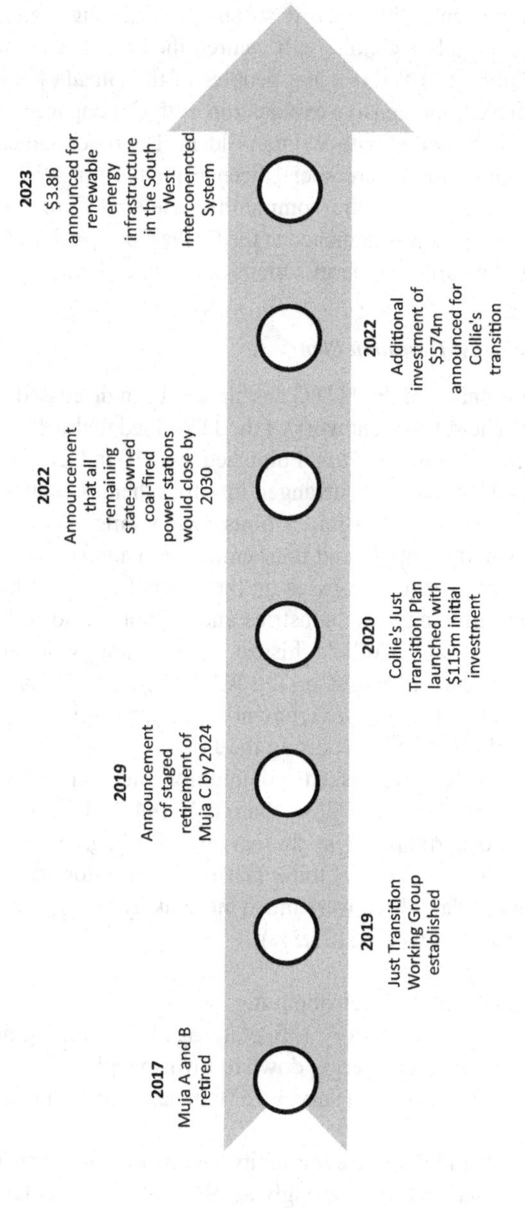

Figure 4.3 Timeline of government decision-making regarding Collie's transition from coal-fired energy production.

members, Climate Justice Union WA and union groups. The report detailed climate-responsive industry opportunities for Collie's transition, including renewable energy, sustainable building materials and recycling renewable technology. It centred the knowledges, wisdoms and responsibilities of Aboriginal peoples of the Gnaala Karla Booja region, with a comprehensive explanation of the Noongar Six Seasons and a section co-written with Wilman Elders. The report called for the WA Government to 'secure social licence for the transition through support for workers and the community' (Beyond Zero Emissions, 2019: 7). The report was launched at the Collie Coal Workers Club and was endorsed by all four energy unions active in Collie.

2020: Collie's Just Transition Plan

The 2019 formation of the JTWG has already been discussed earlier in the chapter. The subsequent work of the JTWG led to the development of *Collie's Just Transition Plan.* Published by DPC in December 2020, the plan has 12 key actions arranged into four themes centred around jobs and the economy. Theme 1 aims to maximise new opportunities for affected workers (and their employers) and support workers to access them. Theme 2 focuses on the diversification of the Collie economy by growing new industries and supporting local business. Theme 3 celebrates Collie's history and future, with economic undertones. Theme 4 focuses on the WA Government's commitment to a just transition through government support, engaging the community and employers, and supporting the JTWG.

In our project, we asked community members about their perspectives of the Plan. Less than half (44.7%) of community members who participated in the survey either agreed or strongly agreed with the statement, 'I think that I am well informed about the Just Transition Plan'. Men were much more likely to agree or strongly agree (56.3%) than women (37.3%).

> I don't really know much about it.
>
> (Survey 135, male, aged 20, working full-time)
>
> Information isn't filtering down to townspeople.
>
> (Survey 76, female, aged 54, unemployed)

Less than half (44.7%) of community members who participated in the survey either agreed or strongly agreed with the statement, 'I feel like I can have my say about the Just Transition Plan' (45% of women, 44% of men). Many community members, particularly those who already experience disadvantage, have not been adequately consulted or engaged in planning Collie's transition. Aboriginal people who participated in our project feel they have been excluded from transition planning discussions and decision-making:

> There's no recognition of consulting with the Aboriginal people at all.
>
> (Workshop 3)

A Traditional Owner said that our project's research workshop was 'the first time ever it involved the Noongar aspect, being asked and involvement in the future of the town' (Workshop 1). Other community members express a desire to be heard, but feel their perspective would not be valued:

> My say generally makes no difference. Business call the shots.
>
> (Survey 96, male, aged 55, working part-time)

Some community members provide suggestions to improve information sharing about the transition, including the 'need to involve local community members and representatives from a variety of cultures' (Survey 130, female, aged 53, working part-time).

Only 35.3% of community members who participated in the survey either agreed or strongly agreed with the statement, 'I am finding the Just Transition Plan to be just (i.e., fair)'. Men were much more likely to agree or strongly agree (43.7%) than women (29.4%). Community members strongly feel that 'just transition is not just about the workers, it's about community' (Workshop 1). However, several community members feel that *Collie's Just Transition Plan* is only about workers in affected industries, and that there are few opportunities for community members to inform and participate in planning Collie's transition.

Some community members feel positively about the Plan's process and implementation. They express that government bodies are engaging with affected workers; that new industries are being planned; that some funding is provided to the community and tourism is being boosted; and that the transition experience in Collie is better than elsewhere in Australia:

> These changes are well overdue but very welcome from me. I see nothing but a positive clean future for the south west and Collie in particular.
>
> (Survey 126, male, aged 54, unemployed)

However, some other community members feel the implementation is not yet delivering results or that it is too fast, with proposed projects that haven't yet come to fruition. Some say there is 'lots of talk and promises' (Survey 44, male, aged 37, working full-time), and 'a lot of fluff talked about and not much happening in town' (Survey 122, female, aged 51, working part-time).

Multiple tranches of state government funding

Substantial state government financial support is a key characteristic of Collie's transition. An initial investment of $115 million announced in 2020 provided funds for the establishment of the Collie Small Grants Program for community grants worth up to $100,000 and the Collie Future Industry Development Fund for grants up to $2 million for companies to invest in Collie. $38 million was also allocated for tourism initiatives in Collie.

In 2022, a $200 million Collie Industrial Transition Fund was established to 'support large-scale industrial projects that can provide long-term, sustainable jobs in Collie' for new and emerging industries (DPC, 2022b). Priority sectors for this funding include green manufacturing or minerals processing, energy-intensive industries, and future clean energy industries. Also in 2022, $300 million was announced to decommission the State-owned coal-fired power stations when they are retired (DPC, 2022c). Finally, in 2023, $16.9 million was announced to support the Collie Jobs and Skills Centre to provide 'free and practical career, training and employment assistance, and is the base for South Regional TAFE's Training Transition Team' (McGowan Labor Government, 2023).

By June 2023, the WA Government had announced more than $662 million in funding to the Collie Transition Fund to support Collie's economic diversification as well as $3.8 billion to 'create new renewable power infrastructure on the South West Interconnected System (SWIS), which powers much of the southern half of WA' (DPC, 2022c). The WA Government asserts that the Collie community will benefit from this investment, which will create 'thousands of jobs' in Collie and the southwest.

What proved easy to achieve and what was hard?

The WA Government's significant financial investments in Collie to support the emergence of alternative industries have produced a number of key achievements so far, particularly investment in skills training and new manufacturing industries. However, initial policies and investments have ignored First Nations peoples and Country and have largely focused on affected industries and jobs rather than the broader community. Even where early successes can be identified, our project showed that there are potential hurdles ahead.

Relatively easy: support and training for affected workers

Provision of support and education for people to pursue alternative employment pathways and for viable career pathways for young people have been relatively easy to achieve, particularly for directly affected workers employed by Synergy. Synergy has developed a *Workforce Transition Program* for workers affected by the transition which provides pathway support, services and resources for employees to achieve individual transition plans to retirement, a new role within Synergy or opportunities outside Synergy. A community member said,

> I admire the way Synergy, for instance, is doing one-on-one, face to face consultation. They are trying to empower their workers, to give them ways they can take back some control over their lives and their future, when it has apparently been snatched away from them.
>
> (Survey 36, female, aged 80, volunteer)

Sustained pressure from unions and workers has also resulted in the WA Government investing in training facilities in the community. The WesTrac Technology Training Centre has been purpose-built to train people in 'the technical skills of operating autonomous equipment for use in the resources sector' (DPC, 2023b). The Government-funded Collie Jobs and Skills Centre opened in April 2023 in Collie Central

Shopping Centre and is available for the entire community. But it is unclear whether this new resource and investment is culturally responsive and will address structural barriers to employment for Aboriginal community members, people with disability, young people entering the workforce and people (predominantly women) with caring responsibilities who cannot work due to lack of childcare.

Relatively easy: promoting economic diversification through tourism

Initiatives to enhance the tourism industry in Collie have also seen relatively quick success. WA and Australian artists were commissioned to contribute to a mural trail with over 40 murals in public places, including an 8000 m^2 mural on the wall of the Wellington Dam. This mural trail is an 'outdoor art gallery' that 'tells stories about the Collie River Valley's living heritage', and meanders through the Collie town to 'encourage visitors to walk around and support local businesses' (DPC, 2023b). $10 million was also invested in developing Collie's Adventure Trails, with 100 km of mountain biking and hiking trails and infrastructure to showcase the region and boost the local tourism economy. The government has funded streetscape revitalisation in Collie, with the repair and upgrade of the façades of 11 historical buildings, and redeveloped Lake Kepwari, a former open cut coal mine site, with recreational infrastructure for swimming, camping and water-skiing. The government also constructed the Kaneang Wiilman Suspension Bridge over *Mardalup* (the Collie River) on the Wiilman Bilya walking trail.

These activities contribute to some goals in Collie's transition plan to showcase Collie's history, change perspectives of Collie, and grow the tourism industry. Community members in our project generally welcome the enhanced infrastructure and beautification of the town, but are concerned that there may be an over-reliance on tourism as a viable economic industry in Collie:

> Don't rely on tourism. It is not inclusive for lower socio-economic community members.
>
> (Survey 118, female, aged 53, working part-time)

To date, no transition strategies enhance or protect the environmental values of *Boodja* (Country), *Bilya* (River) or *Kep* (Water).

Relatively easy with funding: enabling new manufacturing industries

Support for new manufacturing industries has also been quickly forthcoming, supported by Government funding. $2 million was provided for the development of a graphite processing plant, which will 'see manufacturing of graphite products for new technologies including mobile phones, computers, high-pressure sealants and fire-retardant building materials' (DPC, 2021). The first stage of the project is expected to create 40 full-time jobs.

The government has also supported medicinal cannabis company Cannaponics with $2 million for a new 'commercial cultivation, extraction, processing and distribution facility to compete in the rapidly growing medicinal cannabis industry' (DPC, 2023b). Finally, it has invested $13.4 million in the Koolinup Emergency Services Centre in Collie. The Centre is 'a base for an emergency driver training school for DFES staff and emergency service volunteers, and also functions as the State's first regionally-based Level 3 Incident Control Centre during major emergencies' (DPC, 2023b). It is intended that this facility will enable Collie to become an emergency services vehicle manufacturing hub.

Community members see these new industries as vital for Collie's future, but argue that there is still significant work needed to generate long-term new, clean industries that centre *Boodja* and community:

> Great work is going on to attract new industries, but it's not yet as successful as it needs to be… The Collie Futures grants are fantastic and lots of small businesses are getting them and doing good with the money. In the big coal industry workplaces, what is being said to be happening, and what is actually happening at ground level, are two totally different things.
>
> (Survey 93, gender not specified, aged 50, working full-time)

Difficult: promoting an inclusive whole-of-community transition

Community members in our study support these early initiatives to diversify Collie's economy, particularly given that over 1250 people are currently employed in the coal-fired power stations and coal mines:

> Bold vision, attracting and retaining skilled workers. New industry to keep the families local, all of community support to build

> confidence that we are not being left stranded (Survey 40, male, aged 34, working full-time).
>
> Diversification of full time liveable income stream, availability of young families to live in the community, less reliability on welfare (Survey 42, male, aged 42, working full-time).

However, at the same time they express frustration that the transition process so far through the JTWG and CDU has focused primarily on those employed in 'affected industries' rather than a broader community-based transition that centres on the needs of Country and vulnerable community members. Our study showed that the JTWG and CDU have not successfully included or consulted with diverse groups of people in the community, including Wilman Elders and other community leaders, and that as a result diverse lived experiences are insufficiently represented in community planning and decision-making for the transition:

> Ask them. Listen to them. Plan for change.
>
> (Survey 118, female, aged 53, working part-time)

Which key opportunities were missed and with what effect?

Exclusion of Country and First Nations peoples

To date, Collie's transition has missed the opportunity to include Wilman Traditional Owners in transition planning and implementation and incorporate their knowledges, wisdom and cultural responsibilities for *Boodja* (Country), *Bilya* (River), *Kep* (Water), and *Moort* (Family and People). In particular, formal transition processes including the JTWG have not sufficiently supported Elders' access and communication needs to meaningfully participate. Structural marginalisation of First Nations peoples from transition planning further entrenches the deep injustices and inequalities that First Nations peoples already experience, related to areas such as employment, education, health, housing and incarceration. Research demonstrates that First Nations leadership and self-determination in decision-making about land and community also enhances their individual and collective well-being (Gee et al., 2014). As such, this missed opportunity is both one of incorporating the insights that Wilman peoples could bring to transition planning and of beginning to address injustices that have been inflicted upon them.

In our research, community members assert that the transition process should care for *Boodja* through environmental management, land rehabilitation, repairing and cleaning the river system (Survey 108, female, aged 52, unemployed). They call for 'nurturing the microclimate of the Valley' (Survey 82, female, aged 50, working full-time) to 'minimise environmental change' (Survey 64, male, aged 57, working full-time). Wilman Traditional Owners propose that the government invest in cultural ranger programmes that employ Wilman people to care for *Boodja* in culturally-responsive ways. In particular, Elders suggest that in Collie's fight against climate change, local Wilman custodians and knowledges are best placed to support local government and the wider community to care for *Boodja* and *Moort*:

> They've [Wilman peoples] got the knowledge and they could take it back and they'll be heard.
>
> (Workshop 3)

Lack of inclusion of diverse lived experiences in transition planning

Another key missed opportunity to date has been that the community as a whole has been insufficiently engaged in Collie's transition planning or activities. Community members are concerned that the transition will involve significant loss of jobs, with many families and workers leaving town. They assert that this potential population decrease will impact local facilities, services, clubs and opportunities, with reduced funding and service improvement in the community, and decreased in value and sustainability of housing and businesses:

> Community groups will find it hard to maintain members and volunteers (Survey 54, male, aged 56, working full-time).
>
> Loss of medical and support services, loss of education, ie TAFE (Survey 57, female, aged 63, working part-time).
>
> Sports clubs may be even less able to field teams than they are now. Perhaps there will be less need for club facilities and more need for casual recreational facilities (Survey 53, female, aged 63, retired).
>
> Currently our mining companies sponsor a lot of events and areas in town... will the new businesses do this? (Survey 108, female, aged 52, unemployed).
>
> Volunteers will leave town. Shops will close. Services will leave town (Survey 84, female, aged 64, retired).

People who experience disadvantage and marginalisation in Collie have been notably excluded from transition planning, because community members beyond affected workers and the local government are not included in the JTWG and have minimal voice in the transition. Community members urge the JTWG and its implementation of the transition plan to address the needs of the whole community:

> Don't forget about us. Don't leave us behind for the sake of a few votes.
>
> (Survey 13, male, aged 37, working full-time)

Community members want diverse opportunities for the whole community, not just directly impacted coal workers.

Key lessons from this regional transition experience

Our participatory social justice analysis of Collie's transition identifies some clear lessons for other transition initiatives.

Avoid a narrow focus on new industry opportunities

The first lesson is that transition planning which takes a narrow focus on usually male-dominated industries such as manufacturing, construction and emergency services will tend to neglect broader social justice issues that the community wants to see addressed. In Collie's case, this includes the degradation of Country, access to quality housing, social services and physical and mental health support, and access to appropriate childcare and aged care. Community members in our study strongly argue that care industries are central to Collie's transition and can provide low-emitting, meaningful jobs while also addressing inequalities:

> There's a big, missed opportunity in town, of like, transitioning away from being like a fossil fuel working town into more of a family friendly, disability friendly, aged care—because there is a big ageing population.
>
> (Workshop 4)

In order to support employment pathways in these areas, community members call for a range of diverse education and training opportunities to be made available in Collie including providing local Cultural

Ranger training and expanding the local TAFE to include studies in hospitality, nursing, aged care and environment.

To achieve new, clean industries with sustainable jobs for existing and future workers, transitions must foster extensive education and training opportunities that are high quality, accessible, and affordable for community members of all ages and diversities. This might include retraining, one-to-one support, career guidance, and counselling for workers and businesses to diversify. Education and training pathways must be developed and maintained for young people to upskill and actively pursue a career in their community.

Country and First Nations peoples must be central to transition planning

The second lesson is that First Nations peoples must be centred in transition planning. So far, they have largely been excluded from leading, engaging in and contributing to planning Collie's transition. This means that Wilman *Boodja* and the knowledges, rights and responsibilities of Wilman peoples are not centred in the transition plan. In the 26 pages of *Collie's Just Transition Plan*, there are only two references to Aboriginal people: an initial Acknowledgement of Country and a brief mention of a desire to 'maximise opportunities for local small to medium enterprises, including Aboriginal business, through existing and new investment' (DPC, 2020: 21). In our project, community members call for Wilman *Boodja* and her people to be central to transition planning and implementation.

Centring Country and First Nations peoples in transition could help address the myriad social justice and human rights issues that First Nations peoples experience due to ongoing impacts of colonisation, social exclusion and discrimination, and facilitate a programme of reparations. In Collie, Wilman Elders repeatedly emphasise that as Noongar people are the carers for everything, Wilman engagement in the transition process will ensure that First Nations peoples and *Wadjellas* (non-Indigenous peoples) are taken care of. The transition offers an opportunity to systemically address these injustices, and in doing so, enable 'better outcomes for cultural healing' (Survey 119, male, aged 66, retired, Aboriginal) and prevent ongoing racial discrimination towards Aboriginal community members. This is particularly important as Collie's environment continues to degrade due to climate change, which may worsen employment, health and wellbeing indicators for the local Aboriginal population.

Long-term commitment and resourcing from government is essential

The third lesson is that long-term commitment and substantial financial resources from government are instrumental to facilitating transition planning. A key strength of Collie's transition has been the WA Government's funding. This has been possible because a substantial portion of Collie's coal-fired energy industry is government-owned and operated.

However, Collie community members express concern that while transition funding in Collie 'looks big on paper' (Writing workshop), it appears to favour business. There are structural barriers to community groups and non-profit organisations accessing grants to enable community-led jobs creation. A particular barrier is the requirement for matched funding, which affects everyone but particularly First Nations groups. First Nations people suggest that this matched funding requirement should be removed for Wilman applicants in the spirit of reparations: 'The Government should be matching us with what we've lost out on' (Writing workshop).

Community inclusion and participation are the heart of a transition

It is vital that people in all their lived experiences are consistently considered, included and celebrated in a transition process. Transition planners must engage and communicate with local and diverse voices and leaders and foster equitable and inclusive participation in transition planning and decision-making, as well as provide timely and accessible information, communication and resources about the transition process. In our research, community insights highlight several groups of people at risk of being left behind during Collie's transition. These include:

- Workers and sub-contractors who are not directly employed in the affected industries and their families;
- Low income households who have fewer resources to relocate for work;
- Aboriginal community members who are already highly disenfranchised in the local economy, community and decision-making;
- Children and young people, particularly when pursuing local educational and employment opportunities;

- Disabled people and older people, due to the risk of reduced services; and
- Women, non-binary and LGBTQIA+ peoples, due to the policy focus on male-dominated industries.

Our research found that community members want the energy transition to strengthen the Collie community. This relates to preventing population decline, enabling people to remain, work, thrive and age in place, and attract new residents, visitors and businesses to Collie. The transition must maintain community spirit:

> I think that if we have got something of the same spirit that has kept us going for 125 years then we can say we've had a good transition. (Workshop 2)

New industries must enhance Country and community and provide meaningful and secure jobs with decent wages, conditions and support for workers in all their diversities

In our work, Collie community members articulate concerns about the impact of the transition on workers in affected industries, including employees and contractors at the power stations and coal mines, along with other industries such as retail, tourism and services, and local small businesses. Beyond Zero Emissions (2019) reiterates the importance of transition planning for workers in all industries. Collie community members say that new, clean industries must provide secure and stable employment for community members with fair employment conditions that are responsive to diverse individual and family needs:

> A family-friendly, Mum-friendly roster (Workshop 2).
>
> Long term employment with a reasonable income which will allow the town to prosper and continue [to] develop (Survey 55, male, aged 61, working full-time, Aboriginal).

Fundamental to this is ensuring that the transition supports all community members to participate in work, by addressing structural barriers such as caring responsibilities and racism. This includes ensuring equitable access to childcare and addressing barriers to employment for workers with disability or workers caring for people with disability or other needs.

Prioritise social care, affordable housing and health services

Support for vulnerable populations is key to a successful transition. This ranges from infrastructure such as affordable public transport, accessible recreation facilities, and social and affordable housing, to well-funded services in areas such as aged care, physical and mental healthcare, disability support, and childcare. There is a particular deep need for mental health support for affected workers, families and the broader community through the transition process and impacts of climate change, to strengthen community resilience and sustainability.

A just transition must promote climate justice and disaster resilience

Collie community members maintain that economic transition is an opportunity to respond to the changing climate and reduce emissions along with supporting just adaptation and enhancing community resilience to climate change and disasters. New policies, public expenditure, industries and technologies can help facilitate climate justice, while climate resilience also requires coordinated disaster prevention, preparedness, response and recovery from a social justice lens. Climate resilience is an emerging opportunity for industry and community services.

Advocacy for a just transition

Finally, the establishment of a transition plan and significant investment for Collie's industries and affected workers demonstrates the mobilising power of the labour movement in a highly unionised community, and the collective impact of industry and union groups working together. It also demonstrates how advocacy that is overly focused on affected workers can exclude First Nations peoples and marginalised groups. There are real risks that Collie's transition will not be just and will leave some people behind. Collie's experience highlights the importance of including Traditional Owners and people of diverse lived experiences in planning a transition and ensuring that intersectionality and social justice are at the core of advocacy.

Conclusion

This chapter shows how Country, First Nations peoples and community can (and should) be central to a just transition process. Through critically exploring diverse experiences of Collie community members working together to imagine, design, and create just and regenerative zero-carbon economies, this case study extends theories of just transition. The chapter also responds to the research call (Winkler, 2020) for a fuller development of just transition through detailed concrete examples and a community effort. The Collie case study shows that the transition process can and must pull change agents and stakeholders together as a powerful bloc (Simon and Hall, 2002). However, the current process of transition in Collie misses an important and necessary opportunity to dismantle colonialism. Further, the transition is not adequately preparing the Collie community for the intersecting global threats of climate change, economic instability and conflict. Collie's transition policy and investment perpetuates a neoliberal growth model, further marginalising vulnerable peoples and *Boodja.* It is therefore essential to have inclusive participation from all segments of community, led by First Nations Elders, to realise a just transition that leaves no one behind.

References

Beyond Zero Emissions (2019) *Collie at the Crossroads, planning a future beyond coal.* Beyond Zero Emissions Inc. www.bze.org.au/research/report/collie-at-the-crossroads

Clark D (2019) The control of Aboriginal people: 1905 Aborigines Act. *The Carrolup Story.* www.carrolup.info/the-control-of-aboriginal-people-1905-aborigines-act/

DPC (2020) *Collie's just transition plan.* Government of Western Australia. www.wa.gov.au/system/files/2020-12/Collies%20Just%20Transition_09%20December%202020_web.pdf

DPC (2021) *WA's first graphite processing plant to set up in Collie.* Government of Western Australia. www.wa.gov.au/government/announcements/was-first-graphite-processing-plant-set-collie

DPC (2022a) *Collie delivery unit.* Government of Western Australia. www.wa.gov.au/organisation/department-of-the-premier-and-cabinet/collie-delivery-unit

DPC (2022b) *Collie industrial transition fund.* Government of Western Australia. www.wa.gov.au/organisation/department-of-the-premier-and-cabinet/collie-industrial-transition-fund

DPC (2022c) *Collie transition package.* Government of Western Australia. www.wa.gov.au/organisation/department-of-the-premier-and-cabinet/collie-transition-package

DPC (2023a) *Collie just transition.* Government of Western Australia. www.wa.gov.au/organisation/department-of-the-premier-and-cabinet/collie-just-transition

DPC (2023b) *Achievements of Collie's just transition.* Government of Western Australia. www.wa.gov.au/organisation/department-of-the-premier-and-cabinet/achievements-of-collies-just-transition

Find & Connect (2021) *Roelands Native Mission Farm (1938–1975).* www.findandconnect.gov.au/ref/wa/biogs/WE00187b.htm

Gee G, Dudgeon P Schultz C, Hart A, and Kelly K (2014). Aboriginal and Torres Strait Islander social and emotional wellbeing. In P Dudgeon, H Milroy and R Walker, eds., *Working together: Aboriginal and Torres Strait Islander mental health and wellbeing principles and practice* (2nd ed., pp. 55–68). Canberra: Telethon Institute for Child Health Research; Kulunga Research Network.

Jeffers T and Cantatores J (2019) Local jobs and good conditions. *South Western Times.* www.swtimes.com.au/news/south-western-times/local-jobs-and-good-conditions-ng-b881269515z

McGowan Labor Government (2023) *Jobs and skills centre opens to support Collie workers.* Government of Western Australia. www.wa.gov.au/government/media-statements/McGowan-Labor-Government/Jobs-and-Skills-Centre-opens-to-support-Collie-workers-20230406

Mining & Energy WA (n.d.) *Coal.* State Library of Western Australia. https://exhibitions.slwa.wa.gov.au/s/mewa/page/coal

Robertson F and Barrow J (2020) A review of Nyoongar responses to severe climate change and the threat of epidemic disease—lessons from their past. *International Journal of Critical Indigenous Studies* December: 123–138. https://doi.org/10.5204/ijcis.v13i1.1638

Shire of Collie (2022) *Strategic community plan.* www.collie.wa.gov.au/wp-content/uploads/2023/01/Shire-of-Collie-Strategic-Community-Plan-Adopted-December-2022.pdf

Simon R and Hall S (2002) *Gramsci's political thought.* Lawrence & Wishart.

Tattersall A (2022) Alex Cassie – ChangeMaker Chat – Just Transition [Audio podcast episode]. In *ChangeMakers. Stories of people changing the world.* ChangeMakers. https://changemakerspodcast.org/alex-cassie-changemaker-chat-just-transition/

Towie N (2009) Climate change protest outside Muja power plant in Collie. *Perth Now.* www.perthnow.com.au/news/wa/climate-change-protest-outside-muja-power-plant-in-collie-ng-39b46c1aecaf225c5bcab4e7545d9c1a

WA Today (2018) Longest running coal dispute in WA's history ends after 180 days. www.watoday.com.au/national/western-australia/end-of-coal-industry-pay-dispute-looming-as-miners-rally-in-perth-20180212-h0vyhz.html

Winkler H (2020) Towards a theory of just transition: A neo-Gramscian understanding of how to shift development pathways to zero poverty and zero carbon. *Energy Research & Social Science* 70. https://doi.org/10.1016/j.erss.2020.101789

5 Energy transition in the Hunter Valley, New South Wales[1]

Warrick Jordan, Kimberley Crofts and Liam Phelan

Introduction

The Hunter Valley is home to the world's largest coal port, tens of thousands of coal workers, a thriving urban economy in the coastal city of Newcastle and a wealth-generating mining hinterland. A pervasive local optimism has survived several near-death experiences at the hands of a globalised economy since the onset of economic rationalism in the 1980s.

Yet despite this resilience, there are credible fears that the products that have driven the region's prosperity—particularly thermal coal for power generation—will no longer be wanted in a climate-constrained world.

As a result, during the 2019 Australian federal election the Hunter Valley was conscripted as a kind of imagined window into the soul of a restless nation, a prize in the political battle to claim representation of the 'working-family', 'making-things' 'real Australia' (Tranter and Foxwell-Norton, 2021). This contest punctuated a period that has embedded the significance of the global energy transition for the region and its people.

The Hunter is now taking tentative but increasingly purposeful steps towards an energy transition. However, translating principle to action through the fog of fear, sectoral lock-in, identity and politics has been challenging. The need for proactive leadership has become apparent as competing interpretations of values such as fairness, and the hard realities of experienced and projected economic change, influence the scope for action (Evans and Phelan, 2016).

In this chapter, we introduce the key actors, local influences, events and circumstances that have made energy transition a front page

DOI: 10.4324/9781003585343-5

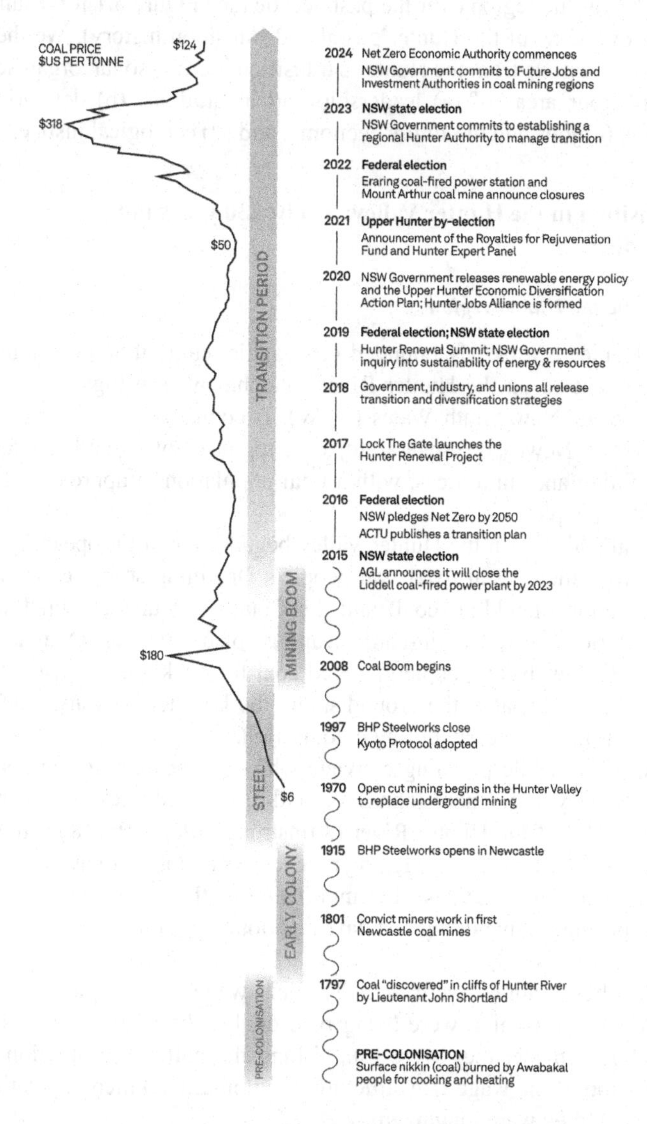

Figure 5.1 Coal and transition history of the Hunter Region, pre-colonisation to 2024.

priority for the region over the past decade (see Figure 5.1 for a summary overview of the Hunter's coal and transition history). We then draw out lessons from the Hunter's transition journey so far, organised around four areas of (a) leadership and institutions, (b) delivering justice, (c) jobs and the future economy, and (d) ecological justice.

Transition in the Hunter Valley: a shift 230 years in the making

Coal and the Hunter region

The Hunter is a clearly bounded geographic, cultural and economic region centred on the Hunter River's catchment (see Figure 5.2) in the state of New South Wales (NSW). It comprises the urban area of Greater Newcastle, several adjacent regional towns and substantial mining and rural areas, with a total population of approximately 750,000 people.

Coal's history in the Hunter Valley began prior to European invasion, with the coastal Awabakal people's Dreaming story describing the creation of nikkin (coal) and the Country around coastal Lake Macquarie named Nikkin-bah, meaning 'place of coal' (Maynard, 2004: 50) Awabakal people gathered coal for cooking, but also held the belief that coal in the ground should be left there (Evans, 2009). This was upended early during colonisation.

In 1797, while pursuing convicts who had escaped up the coast from Sydney, Lieutenant John Shortland discovered thick coal seams in the cliffs of the Hunter River (Armstrong, 1983). By 1801, there were three trained miners digging nine tonnes of coal per day. Though these men were convicts, the importance of their work gave them extra privileges, including rest days and double rations:

> As these rations were actually their wages, these pioneers of Australian mining were being paid double the rates of unskilled labour. In fact they were establishing the collier's profession at the top of the wage rate structure of colonial workmen, a position which they were long to enjoy.
>
> (Turner, 1982: 18–19)

From these beginnings, coal has permeated the life of the region—forging its working-class character, driving its economy, influencing

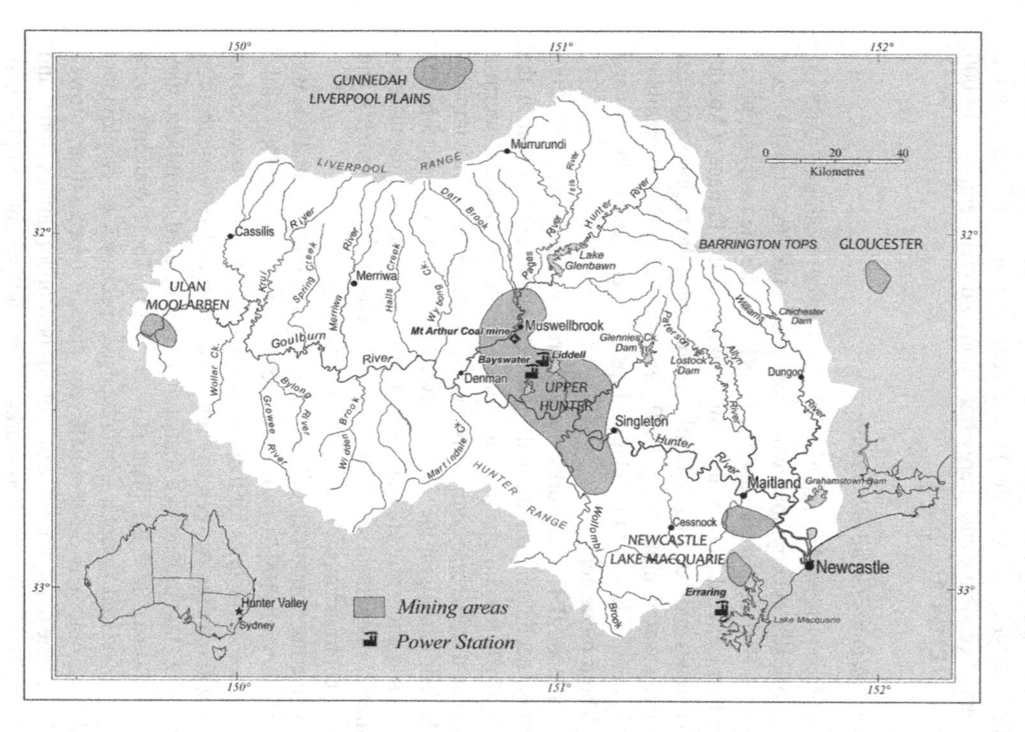

Figure 5.2 Mining areas in and near the Hunter River catchment (Rey-Lescure, 2024). Note Mount Arthur Coal Mine near Muswellbrook, the now closed Liddell Power Station on the shores of the artificial Lake Liddell in the centre of the map and Eraring Power Station on the shores of Lake Macquarie at bottom right.

its political economy, scarring its landscape and waters and supporting the livelihoods of many tens of thousands (Wells, 1950). This influence persists to the present day. Export coal is the largest direct contributor to the region's Gross Regional Product. It contributes $22 billion per year, 15% of the Hunter economy, and—critically in the context of economic and energy transitions—directly employs 14,000 people (REMPLAN, 2023).

A recent history of structural change

The Hunter is defined by experiences of industrial transition. Newcastle is historically synonymous with steel production in Australia, ever since the BHP steelworks opened in 1915 to take the ready supply of metallurgical coal from nearby mines to power the furnaces (Abbott, 1997; Jones and Tee, 2017). Until the late 1990s, the region's political economy was dominated by BHP, several other coal and manufacturing companies, and the relationships between those industries and the labour movement.

While the BHP steelworks enjoyed great success through the 20th century, declining demand and increased global competition in the early 1980s led the company to reduce employment and modernise equipment to increase competitiveness. This culminated in a company-wide restructuring in the 1990s (Lewer, 2013), partly assisted by the national Labor Government's industry programme, guaranteeing domestic markets in exchange for productivity improvements and labour commitments (Abbott, 1997).

A steering committee of executive and union representatives was formed in early 1997 to guide the Newcastle restructure. One of the committee's principles for that transition was that no employee should leave without dignity, yet many felt betrayed when BHP announced in April 1997 that instead of a transition to restructured operations, they would instead close the plant (Lewer, 2013). These concerns were partly alleviated through comprehensive packages of employment assistance and a two-and-a-half-year closure process that allowed staff the time to retrain, retire or look for new opportunities (Jones and Tee, 2017). However, 4000 workers were left without employment (Lewer, 2013) and the region experienced several years of challenging economic times when efforts to attract large-scale substitute employment failed. In truth, the region's subsequent renaissance following the closure of the steelworks was largely due to the good fortune of a coal export boom and broader shifts towards a service economy.

The demise of the steelworks had enduring implications for the region's capacity to organise responses to transition. The move from an industrial company town run by Australia's largest company to a more diversified economy disrupted the coherence and influence of regional business, labour and political power structures. The effects of this fragmentation persist to the present day.

2001–2014: the mining boom and climate change

New external factors began to shape the Hunter's trajectory in the first decade of the 2000s. By 2008, coal was fetching US$180 per tonne, a sixfold increase in the price in five years, driven by a strong global economy and unprecedented growth in demand from China (Brett, 2020). As a source of high-quality coal with good infrastructure and low sovereign risk, the Hunter was well placed to meet demand. The associated influx of foreign capital fundamentally reshaped its economic landscape, bringing well-paid employment to tens of thousands to build and operate export-oriented mines and enabling infrastructure.

But like other resource booms, the Hunter's coal boom also brought with it inflation and labour shortages in other industries as workers flocked to high-paying mining jobs, as well as structural vulnerability due to over-reliance on a single sector. The increasingly visible social and environmental costs of mining led to community campaigns to return royalties to the region and address landscape and pollution impacts. The boom also raised the profile of the region in national debates on climate policy and action, as local protests highlighted the fact that the region hosted the world's largest coal port and so was of global relevance to climate change.

When the boom ended in 2013, it was a serious regional economic shock which brought a sharp rise in unemployment. Oversupply and weakening demand combined with the end of the labour-intensive construction phase for greenfield mine development. Thousands of workers were laid off and marginal operations closed. The visible consequences of the bust and the lack of government support to manage its impacts increased awareness in the Hunter of its economic vulnerability and led to interest in diversification, particularly in the most affected mining communities of the Upper Hunter. However, the demand for export coal stabilised and it remains significant to the present day.

2015–2024: key events, decisions and actions in the Hunter's energy transition

Contemporary energy transition initiatives must be understood in the context of this longer history. The current energy transition process arguably began in 2015 but intensified considerably between 2020 and 2022. Four distinct sets of key events have shaped the transition conversation in the Hunter in this period:

1. The announced closure in 2015 of the Liddell coal-fired power station;
2. The 2019 federal election;
3. The state government acknowledgement of the reality and need for transition from late 2020; and
4. Three events from the first half of 2022: the announced closure of the region's largest power station (Eraring) and coal mine (Mount Arthur) and the election of the federal Albanese Labor government.

In early 2015, energy company AGL had announced plans to close Liddell Power Station. The initial response was low key, giving rise to a company-state-university coordination effort with the Hunter Energy Transition Alliance. But this decision became the event that lit the fuse on the transition issue as Liddell rapidly found itself at the centre of a national debate when the federal government sought to ensure AGL kept the power station open.

In the Hunter, tangible events like power station closures began to broaden acceptance of the idea that a proactive response to transition was needed. Economic diversification had been of local interest for many years, but by early 2016, following the announced closure of Liddell, regional development and transition become a more prominent issue. This included proposals from local government for a $30 million Upper Hunter Economic Development Corporation.

Later in 2016 the transition away from coal was thrown further into the spotlight when the owner of Hazelwood—the largest power station in Victoria's Latrobe Valley—announced that it would imminently close the station down (see Chapter 3). This prompted advocacy groups to step forward at multiple levels. The Australian Council of Trade Unions (ACTU) published a transition plan focusing on the need to minimise the impact of unplanned closures on regional communities (ACTU, 2016); the Mining and Energy Union released a detailed study informed by past failures and successes in Germany; the United States and Newcastle (Sheldon et al., 2018); and environmental organisation Lock the Gate launched the Hunter Renewal Project (Hunter Renewal, 2023).

By mid-2018, key actors including the NSW Government, local councils and the Port of Newcastle had released diversification strategies that directly recognised the economic challenges of coal export dependence (Green, 2018; NSW Government, 2018). At the same time, Hunter mining communities were becoming increasingly concerned about the prospect of climate policy constraints on coal exports—concerns that exploded with significant political force during the 2019 federal election.

The influence of climate issues on the tightly contested 2019 federal election has been extensively documented (Cameron and McAllister, 2020; Tranter and Foxwell-Norton, 2021). A prominent issue was advocacy from environmentalists for federal intervention to prevent a large new coal mine in Central Queensland by the Indian company Adani, including a protest 'convoy' (Colvin, 2020). Sentiment on climate and mining issues became further polarised, including in regional mining areas in Queensland and NSW. In the electorate covering the key mining areas of the Hunter, a coal miner representing the populist One Nation party made an unprecedented attempt to unseat the incumbent Labor MP.

These conditions had significant local and national effects, revealing vulnerabilities in both the major (Labor and Liberal-National) political parties about how to manage the different views of urban and regional constituencies on climate issues. Discussions about transition were reframed in some affected communities, including parts of the Hunter, as being solely equated with imposed closures of coal assets, rather than attempts to proactively respond to inevitable shifts in the global energy economy.

During this volatile period of political conflict and paralysis, the idea of transition was sustained largely by a set of local stakeholders. For example, in February 2019, local environment group Hunter Renewal hosted ex-coal miners from the US state of Kentucky to relate their experience of a region failing to prepare for inevitable change. In July 2019, local independent state MP Greg Piper and other independents led an inquiry on renewable energy and the economic consequences of transition (Parliament of NSW, 2019).

In late 2020, the Hunter Joint Organisation of Councils, a statutory representative body for the region's local governments, began advocating for a regional transition body. The Hunter Jobs Alliance, an alliance between local unions and environmental organisations, was formed around this time. It also advocated for a regional transition authority, including worker support and regional development investment (Hunter Jobs Alliance, 2021a). Some major regional industry actors including coal miner BHP and energy company AGL also continued to lay groundwork for transition activities that reflected their corporate strategies.

Towards the end of 2020, a series of state government actions began to suggest greater acknowledgement of the need for action on transition, as well as an implied recognition that the community was seeking this direction. A substantial renewable energy policy was put in place by the conservative state government, supported by the Labor opposition (NSW Government, 2020a). NSW Government documents such as the Intergenerational Report began to acknowledge that change was coming, including in relation to coal royalties (NSW Government, 2021). Of particular significance was a Strategic Statement on Coal Mining and Exploration in NSW recognising that 'the transition to new energy sources is a long-term economic change that will continue to reshape our regional communities that currently rely on the export coal industry' (NSW Government, 2020b).

A state by-election in the key mining seat of Upper Hunter in May 2021 was a significant policy inflection point. While both the National (rural conservative) and Labor parties expressed strong support for mining, both also developed policies designed to respond to rising community concern about economic change along with vigorous campaigning by local and out-of-region actors seeking to progress singularly pro-mining or climate focused policies. These policy proposals included coal royalty reinvestment funds and repurposing of coal mines as employment precincts.

The Nationals (the junior partner in the conservative state Coalition government) retained the seat, bringing with them a commitment to implement the Royalties for Rejuvenation programme. This programme responded to proposals to set aside some coal royalties for transition activities and to establish participatory governance. This programme established a $25 million annual fund to be allocated to four NSW coal mining regions including the Hunter, guided by stakeholder 'Expert Panels' (NSW Government, n.d.).

Transition became an even more pervasive public conversation from early 2022 after Origin Energy announced it planned to close Australia's largest power station at Eraring in Lake Macquarie by 2025, seven years earlier than previously announced (although it is important to note that the closure date has subsequently been held up by delays in renewable energy developments and electricity system challenges). In 2022, BHP also announced that it had failed to find a buyer for the country's largest coal mine, Mount Arthur in the Upper Hunter, and that it would therefore be closing the mine in 2030 (Kelly, 2022).

The election of the federal Labor Government in 2022, including a former coal miner to the seat of the Hunter, marked the return of federal efforts to balance emissions reductions and the value of mining and industry to regional Australia, through a national emissions reduction scheme, regional industry policy and the creation of a national Net Zero Authority. This was followed in March 2023 with the new state Labor government committing to establishing a regional Hunter Authority to manage transition (Kelly, 2023).

While federal and particularly state government transition efforts have taken considerable time to be put in place, for example delays in the implementation of the Royalties for Rejuvenation programme, at the time of writing several key transition policy initiatives were being advanced. These include the establishment of the Australian Government's Net Zero Economy Authority; the announcement of a national Future Made in Australia industry plan focused on the net zero transformation; the NSW Government's release of a proposed design for Future Jobs and Investment Authorities, including one in the Hunter, focusing on coal power and mining transition (NSW Government, 2024); and a NSW Government-led parliamentary inquiry into the reuse of mining lands (Parliament of NSW, 2024).

Challenges in promoting transition in the Hunter

Very little of the Hunter's transition process to-date has been easy. Understanding the stress points and fractures of the region's political economy provides lessons to other regions seeking to navigate the energy transition. Three related issues stand out. These are economic and political economic 'lock in', decline of comparative economic advantages and a political paralysis deriving from social fragmentation.

The natural riches seen through the coal export boom have created a regional political economy which is not incentivised to explore new opportunities or prioritise the difficult work of intentional economic diversification. In a pattern shared with many regions, this political-economic lock-in has made it difficult for communities and businesses in the Hunter to recognise impending irreversible structural economic change (Moretti, 2012). This has been compounded by a local tendency towards incremental managerialism, in which business and political leaders are incentivised to make small adjustments to maximise the benefits of external demand drivers rather than relying on local innovation. This lock-in has slowed the process of diversification and made it hard to convince locals that transition is inevitable.

Underlying these failures to innovate and plan ahead are stark realities that risk turning local strengths into weaknesses. While the Hunter's wealth has come from global demand for its coal this economic strength also obscures the challenges of creating ongoing economic activity when that demand disappears. Ultimately, the region is a capital and price taker on the global periphery. It lacks non-coal mineral resources, is exposed to the ongoing impact of globalisation on the competitiveness of domestic manufacturing and lacks tradeable knowledge industries. While the Hunter is not without strengths, confidence and opportunities, the disruptive up-ending of structural economic certainties that have prevailed for generations is a material challenge to managing the transition, and a direct concern for local households.

A further challenge has been the erosion of community cohesion and social capital. In some measure, the Hunter's climate and transition contests are local manifestations of wider political and social fragmentation across the world, for example through education, location and income cleavages (Gethin et al., 2022). Disagreement on who in the community deserves help during change, and the level of government intervention that should be permitted on environmental and economic issues (Productivity Commission, 2017) also reflects broader Australian cultural shifts, such as the embrace of individual aspiration as a replacement for widely shared egalitarian values of 'fairness' (Dyrenfurth, 2007).

Beyond these broader trends, transition processes have in some cases led to community fragmentation in the Hunter, though this has been substantially moderated by proactive local efforts. For example, previously strong community bonds have eroded as divisions between those relatively more supportive of climate action and those prioritising local employment have grown. This 'jobs versus the environment' narrative has unhelpfully dominated energy transition discourse (Instone, 2015).

The resulting polarised political context has led to bouts of paralysis grounded in political point-scoring and policy risk-aversion. However, this politicisation obscures and misrepresents a more considered and widespread community view that acknowledges transition and respects both economic and environmental imperatives, for example as revealed in a study of Upper Hunter residents in early 2020 (Colvin and Przybyszewski, 2022). While this does not mean those different interests are easily reconciled, it does point to prospects of a more deliberative and cohesive approach than that suggested by actors hell-bent on pursuing exploitative political polarisation.

Addressing these constraints in acknowledging change, adjusting to new economic challenges, and navigating through polarisation have been substantially aided by proactive efforts to engage with the community to identify actual sentiment and to discuss appetite for addressing change. As these discussions have become more commonplace, evidence has been provided that the transition is recognised even if not universally welcomed, and that communities have confidence in the Hunter's future and want a coherent response to ensure future prosperity (Institute for Regional Futures, 2023, 2024).

Moreover, communities bring a commonsense practicality to the table that seeks to move beyond transition discussions to tangible actions. For example, community workshops held across the Hunter in late 2020 revealed strong appetite for clear planning, community participation and practical industry outcomes (Hunter Renewal and Hunter Jobs Alliance, 2021), as the following comments from participants illustrate:

> If we don't have adequate planning as we transition out of fossil fuels there is a very real prospect that some people will be left high and dry, not just people who work in the fossil fuel industry, it'll be entire communities.

> The most important thing is involving the local community in designing the transition. Unless you take the locals with you on the journey, so that they own the changes, it will not be successful.
>
> We should be starting on planning for decarbonising industry. We have to do something about it everywhere, we have to start thinking about it now.
>
> (Hunter Community Workshop Participants)

Key learnings from this regional transition experience

The Hunter's transition is far from complete, but the region's experience to date provides some instructive lessons. While place-specific, these learnings may resonate with other communities in similar situations. Noting that similar descriptions have appeared in the scholarly literature (e.g. Evans and Phelan, 2016; Wiseman et al., 2017; Harrahill and Douglas, 2019; Weller et al., 2024), we have organised these lessons across into four themes:

a. Leadership and institutions
b. Delivering justice
c. Jobs and the future economy
d. Ecological justice

a. Leadership and institutions

A key lesson from the Hunter's experience to date is that leadership is required to navigate a path through the transition, and that this leadership must be accompanied by institutions that are fit-for-purpose in tackling the long-term, structural nature of change.

In some respects, the type of leadership required conforms to basic requirements of good public policymaking—clear problem identification, deliberative policy development, a focus on practical implementation, and ensuring there is sufficient key stakeholder and community acceptance to ensure a policy can survive over time (Luetjens et al., 2019).

However, policymaking of this sort has proven highly challenging to employ in the Hunter because of economic lock-in, structural vulnerabilities, and political polarisation, particularly at key moments, such as plant closures and elections. Many of the key regional leaders—members of parliament, organised labour leaders, captains of industry, local mayors, and the like—are directly connected to constituents who are vulnerable to change, concerned about accessible economic alternatives, and often sceptical about the motives and interests of decision-makers.

It has taken considerable time for these regional leaders to conclude that change is inevitable, understand the nature of that change and how to respond, and find ways to communicate these new and sometimes uncomfortable realities with their constituencies in a way that does not aggravate fears or undermine the prospects of putting in place effective transition actions.

Compounding this leadership challenge is the lack of regional institutional capacity to implement the required actions. The Hunter as a region is not a political or administrative entity where someone is easily identifiable as responsible for guiding transition. Most of the decision-making and implementation capacity as well as the funds to underpin transition are located in state and federal capital cities and in boardrooms of national and international businesses, rather than in the region itself.

Australian regional governance tends towards centralisation in higher levels of government and away from delegated resourcing and decision making authority at regional level (Pugalis and Gray, 2016). Given the scale of transition, much of the action needed will invariably be driven from outside the region. However, recognition of accelerating transition in the Hunter has resulted in a resurgence of advocacy for regionally empowered agencies—transition 'authorities' or structured collaborations—that bring together different actors, develop clear plans and create specialised capacity to address practical needs of the transition (Hunter Jobs Alliance, 2021b; Hunter Joint Organisation of Councils, n.d.). As one community member noted during a regional workshop:

> There needs to be better coordination, but it should come from the community, through council, and then up to government for support.
>
> (Hunter Renewal and Hunter Jobs Alliance, 2021)

While still in commencement phase, the new national Net Zero Economy Authority and a commitment to a state-based Hunter Future Jobs and Investment Authority (NSW Government, 2024) have the potential to respond to these needs.

b. Delivering justice

Transitions are defined by their success in supporting affected people through change in a manner that prevents their exposure to harms and maximises their prospects of meeting their future needs and aspirations (Productivity Commission, 2017). Transitions are also complex and contested processes which require the expertise and participation of the widest possible range of stakeholders across social, economic, policy and environmental areas. The questions of who speaks for a region and who can participate are critical in ensuring those who are exposed to change are supported. Governments therefore must play a central role in ensuring marginalised groups have access to participation.

In the Hunter, advocacy organisations have been careful to create expectations that workers and local representatives should have a seat at the table and that structures created to address transition should adequately combine representation, technical knowledge and local experience (Hunter Jobs Alliance, 2021b). As observed in other regions, technocratic approaches without representation will not be accepted as legitimate by communities (Weller, 2017; Weller and Tierney, 2018).

While governments at all levels come to grips with the idea that participation matters, local actors have sought to fill the participation vacuum with a minor industry of public forums, conferences and conversations (e.g. Phelan and Crofts, 2022). In tandem, advocates—particularly from local government, unions and environment groups—have identified that conventional consultative structures are failing to deliver meaningful participation or results. While these traditional consultative approaches often result in the formation of committees and panels, they are generally limited in effectiveness as they lack permanence, decision-making powers and meaningful resources.

Promisingly, regional, state and national advocacy for more coherent and better resourced governance strategies appears to be bearing some fruit (ACTU, 2016; Hunter Jobs Alliance, 2021b). Emerging institutional innovations at the national (e.g. Net Zero Economy Authority) and state levels (e.g. Expert Panels, Future Jobs and Investment Authorities) speak directly to the need for genuine representation and the need to find and build the practical transition-response expertise of people in the Hunter and places like it.

The test will be whether new authority models deliver on this promise. To be successful, these new institutions must deliver both procedural and distributive justice. The experience from the Hunter is that processes and decision making must be perceived as fair, and that outcomes must be effective and targeted to those affected and in need.

Bridging the gap between discussion and action has proved challenging to date. As one senior public servant wryly noted, the outcome of lots of talking but little action is that the Hunter is 'rich in strategic plans' (personal communication). More coherent capacity will be required for a just transition away from export coal. The coordinated response to the closure of the BHP steel works in Newcastle offers some clues. Training and labour market navigation services delivered by BHP at the time of the steelworks closure had notable success (Lewer, 2015). Other practical actions such as subsidised training for new and sustainable industries, worker transfer schemes, and high-quality labour market services for affected workers have all been highlighted by labour unions and community organisations as being crucial to a fair energy transition in the Hunter (Sheldon et al., 2018; Hunter Jobs Alliance, 2021b). The research being undertaken to inform the emerging transition architecture at state and federal level appears to be informed by and responding to these experiences and proposals for specific worker support activities (Department of Regional NSW, 2024; Net Zero Economy Agency, 2024).

c. *Jobs and the future economy*

Establishing the expectations and means to treat workers fairly through change is half the battle. The other half of the economic transition equation—finding ways to replace economic activity and jobs—is incredibly difficult.

As political conflict has dissipated, early-stage power plant and mine closures have been announced, and awareness of competition for new industries has built, community pressure to 'get on with it' has grown. There is a fundamental challenge in reconciling community expectations for 'traditional' blue collar work with the realities of a modern service based economy where such work is proportionally diminishing. While the mining sector has provided a bulwark against this shift in the Hunter region, as the broader energy transition accelerates over time structural shifts will become more locally apparent. Accepting structurally lower prosperity, however, is not a feasible economic or social strategy (Martin, 2012; Rodríguez-Pose, 2018) so alternatives must be found.

As of 2024, the Hunter is still doing well economically from exporting coal. However, it is becoming increasingly clear that this economic activity is finite and likely to come to a close over the next generation. More sophisticated regional development and investment attraction tools are needed to match this more realistic appraisal of global trends. There is a clear risk for the Hunter that a type of regional development illusion is emerging, where the language and paraphernalia of government intervention multiply in response to community pressure, but no genuinely substantive measures eventuate, as has been observed in other changing industrial economies (Breathnach, 2010).

Fortunately, there are signs of improved approaches emerging. One critical shift has been a renewed government emphasis on regional industry policy that has made the Hunter a national target for job-generating public investments in the clean energy economy. While the effectiveness of such investments at both national and regional levels are yet to be seen—particularly in the context of the global rise of equivalent public investments such as the *Inflation Reduction Act* in the United States, the European Green Deal and other competing initiatives—the substantial public investment infrastructure being assembled is encouraging.

The Hunter is one of the priority locations for the deployment of modernised industry development efforts, particularly under the framework of the Australian Government's Future Made in Australia policy agenda (Prime Minister of Australia, 2024a). Examples include decarbonisation investments from the public Clean Energy Finance Corporation and the climate focused Powering the Regions Fund, the prioritisation of clean energy and manufacturing through the new National Reconstruction Fund, and investment attraction through the Net Zero Economy Authority and industry programmes such as the Solar Sunshot programme (Prime Minister of Australia, 2024b).

Some of the region's largest industrial employers and largest carbon emitters are proactively investing in emissions reduction technology. This includes commitments and major investment in renewable energy to decarbonise the Tomago Aluminium smelter—Australia's largest energy user—and decarbonisation and hydrogen production investments at the region's ammonium nitrate production facility (Orica, 2023). These investments send a clear signal that heavy industry employment can be maintained in the region in an emissions-constrained future.

The other encouraging shift is the reinvigoration of the Hunter's focus on locally driven innovation. When the steelworks were winding down, the Hunter pioneered practical examples of what is now described as 'place-based development'—combining local and expert knowledge in resourced institutions that identify untapped growth opportunities (Rodríguez-Pose, 2018). Several not-for-profits formed in the Hunter during the manufacturing shocks of the 1980s and 1990s, specialising in small business establishment, supply chain diversification, firm clustering, and small and medium enterprise development. Organisations such as HunterNet are now refocusing this strong knowledge base on contemporary transition needs, for example developing supply chains for the emerging offshore wind and hydrogen industries.

Taken in sum, whether focused on assuring the future of emissions-exposed industries with decarbonisation potential, seeking a share of global clean energy industry opportunities, or adjusting to structurally changed economies by creating 'lots of little industries everywhere' (Hunter Renewal and Hunter Jobs Alliance, 2021), there is recognition in the Hunter that more proactive efforts are required. Whether these efforts are effective, sufficient or suitably timed to generate jobs and economic opportunity in the face of energy, technology, climate and broader structural economic trends will be a critical determinant of whether the Hunter's transition is ultimately a fair and effective one.

d. Ecological justice

The acceleration of coal mining over the past few decades has transformed the landscape of the Hunter, with damage that is extensive and largely irreversible. Despite evolving policy responses, rehabilitation science and stakeholder pressure (Unger et al., 2020), conservation and environmental concerns still take a back seat to extraction.

A transition from coal mining across large parts of the landscape presents an opportunity for a revised approach (Hunter Renewal, 2023). The Hunter has a unique biogeography, with overlapping zones of classically Australian dry eucalypt forest, near-outback habitats reaching in from the west, and fertile river plains and forests (Bryant and Krosch, 2016). It is damaged but can be restored (Drinan, 2022). Conversely, failing to deliver ecological justice will undermine the broader prospects of a just transition by unnecessarily alienating a significant section of the community who are concerned about the mining industry's landscape impacts and encouraged by the prospects of long-term landscape restoration.

Global mining giants occupy large chunks of historically and recently damaged ecosystems, enough to join up surviving remnants and fill the gaps with appropriate conservation investments. Local expertise in repairing mining impacted landscapes is substantial and techniques in mine restoration are improving. Australia hosts some of the world's most innovative and collaborative land restoration initiatives, including the Great Eastern Ranges initiative traversing the Hunter (Pulsford et al., 2013). With sufficient intent, the transition could also offer significant new opportunities for First Nations people to care for their lands while finding employment and business opportunities in mine restoration.

While the ecological pieces can never be put together the same way again, there is an opportunity to create a restorative, and potentially world leading, environmental outcome (Crofts and Phelan, 2023). Addressing environmental policy failures requires the same leadership and institutional innovation demanded by the economic transition. The industry shift underway represents a generational chance to reset the relationship between the people of the Hunter region and their landscape, as a critical component in ensuring a truly just transition.

Conclusion

Delivering an effective, socially just and ecologically sensitive energy transition in a coal-dependent region is no mean feat in a context of existential fears and febrile politics.

Energy transition in the Hunter Valley has nevertheless begun. Two decades of inertia have given way to a growing awareness of and action for change. The risk at this point is that the region remains embroiled in political contest and has its view of the future obscured by current prosperity, and consequently fails to act until impacted by an economic crisis that it is too late to do anything meaningful about. The view from 2024 suggests it could go one of three ways.

First, there are green shoots of community recognition, regional leadership, and early efforts at institutional and industry responses. There is promise and momentum, but the challenge is to convert this into practical and concerted action. The alternatives are less palatable.

The second option is that the Hunter's transition could fall into step with the historical, prevailing Australian approach to regional adjustment. That is, to partly acknowledge future change, but to prioritise immediate concerns and to implement half-measures in the hope that the winds of change are manageable when they arrive in earnest. History suggests this is a high-risk strategy, both for those working in the mining and power industries and for the vulnerable members of the broader community that will be impacted by economic change.

The third option is that the region and its leaders simply fail to act until a genuine crisis emerges. Fortunately, this appears to be an increasingly indefensible option in the public discussion, but on a long transition pathway with unpredictable future events, it is impossible to rule out.

The Hunter pursuing a dedicated and proactive transition response is possible, but the effectiveness, scale, durability and community acceptance of the approaches now being considered and established will be pivotal. There is significant reason for optimism, but current trends suggest that a partial, sub-optimal transition response is still the most likely trajectory for the Hunter.

In common with other regions, the Hunter will continue to wrestle with the challenge of transition and the questions of responsibility, justice and action that go with it. How well it answers those questions will go a long way towards determining the well-being of the place, the prosperity of its households and its role in the world's movement towards a more sustainable future.

Note

1 Our thanks to John Maynard for helpful guidance on Awabakal people's pre-invasion use of coal.

References

Abbott MJ (1997) Debate the closure of a steelworks and the limits to workplace reform. *Industry and Innovation* 4(2): 259–275.

ACTU [Australian Council of Trade Unions] (2016) *Sharing the challenges and opportunities of a clean energy economy: A Just Transition for coal-fired electricity sector workers and communities*. www.actu.org.au/our-work/policy-issues/actu-policy-discussion-paper-a-just-transition-for-coal-fired-electricity-sector-workers-and-communities

Armstrong J (ed.) (1983) *Shaping the Hunter: The engineering heritage.* Newcastle: The Institution of Engineers, Australia.

Breathnach P (2010) From spatial Keynesianism to post-Fordist neoliberalism: Emerging contradictions in the spatiality of the Irish State. *Antipode* 42(5): 1180–1199.

Brett J (2020) The Coal Curse: Resources, climate and Australia's future. *The Monthly* 78.

Bryant LM and Krosch MN (2016) Lines in the land: A review of evidence for eastern Australia's major biogeographical barriers to closed forest taxa. *Biological Journal of the Linnean Society* 119(2): 238–264.

Cameron S and McAllister I (2020) Policies and performance in the 2019 Australian federal election. *Australian Journal of Political Science* 55(3): 239–256.

Colvin RM (2020) Social identity in the energy transition: An analysis of the "Stop Adani Convoy" to explore social-political conflict in Australia. *Energy Research and Social Science* 66. https://doi.org/10.1016/j.erss.2020.101492

Colvin RM and Przybyszewski E (2022) Local residents' policy preferences in an energy contested region—The Upper Hunter, Australia. *Energy Policy* 162: 112776.

Crofts K and Phelan L (2023) 'We need to restore the land': As coal mines close, here's a community blueprint to sustain the Hunter Valley. *The Conversation*, 8 February. https://theconversation.com/we-need-to-restore-the-land-as-coal-mines-close-heres-a-community-blueprint-to-sustain-the-hunter-valley-198792

Department of Regional NSW (2024) *Future Jobs and Investment Authorities, Issues paper*. www.nsw.gov.au/sites/default/files/noindex/2024-05/FJIA-issues-paper.pdf

Drinan J (2022) *The Sacrificial Valley: Coal's legacy to the Hunter*. Arcadia, NSW: Bad Apple Press.

Dyrenfurth N (2007) John Howard's hegemony of values: The politics of 'mateship' in the Howard decade. *Australian Journal of Political Science* 42(2): 211–230.

Evans G (2009) *A just transition to sustainability in a climate change hot spot: The Hunter Valley, Australia*. Doctoral Thesis, University of Newcastle, Australia.

Evans G and Phelan L (2016) Transition to a post-carbon society: Linking environmental justice and just transition discourses. *Energy Policy* 99: 329–339.

Gethin A, Martínez-Toledano C and Piketty T (2022) Brahmin left versus merchant right: Changing political cleavages in 21 Western Democracies, 1948–2020. *The Quarterly Journal of Economics* 137(1): 1–48.

Green R (2018) Why we need a world-class container port. *Newcastle Herald*. www.newcastleherald.com.au/story/5547928/why-we-need-a-world-class-container-port/

Harrahill K and Douglas O (2019) Framework development for 'just transition' in coal producing jurisdictions. *Energy Policy* 134. https://doi.org/10.1016/j.enpol.2019.110990

Hunter Jobs Alliance (2021a) *No regrets: Planning for economic change in the Hunter.* https://hunterjobsalliance.org.au/wp-content/uploads/Hunteralliance_planningforeconomicchange_hi.pdf

Hunter Jobs Alliance (2021b). *Building for the future: A 'Hunter Valley Authority' to secure our region's future.* https://hunterjobsalliance.org.au/wp-content/uploads/HJA2021_BuildingfortheFuture_AHunterValleyAuthority_lores.pdf

Hunter Joint Organisation of Councils (n.d.) *Hunter 2050.* www.hunterjo.com.au/resources/hunter-2050/

Hunter Renewal (2023). *After the coal rush, the clean up: A community blueprint to restore the Hunter.* https://assets.nationbuilder.com/lockthegate/pages/8176/attachments/original/1690764718/Blueprint_final_1.pdf

Hunter Renewal and Hunter Jobs Alliance (2021) *Future-proofing the Hunter, voices from our community.* https://hunterjobsalliance.org.au/wp-content/uploads/SINGLE_-_Future-proofing_the_Hunter_report_final.pdf

Institute for Regional Futures (2023) *The Hunter matters, regional insights.* November 2023. https://nova.newcastle.edu.au/vital/access/manager/Repository/uon:53233

Institute for Regional Futures (2024) *Hunter horizons: Navigating the future of work and workplaces in our region.* June 2024. https://nova.newcastle.edu.au/vital/access/manager/Repository/uon:55424

Instone L (2015) Risking attachment in the Anthropocene. In: K Gibson, D Bird Rose and R Fincher, eds., *Manifesto for living in the Anthropocene.* Brooklyn, NY: Punctum Books.

Jones S and Tee C (2017) *Experiences of structural change.* Australian Government.

Kelly M (2022) BHP announces that Mt Arthur coal mine will close in 2030 after it fails to find a buyer. *Newcastle Herald,* 16 June. www.newcastleherald.com.au/story/7782614/bhp-commits-to-consultation-for-post-mining-future/

Kelly M (2023) NSW Labor commits to the establishment of a Hunter clean energy transition authority. *Newcastle Herald,* 15 March. www.newcastleherald.com.au/story/8122616/labor-announces-clean-energy-transition-authority-plan/

Lewer J (2013) Employee involvement and participation under extreme conditions: The Newcastle steelworks case. *Journal of Industrial Relations* 55(4): 640–656. https://doi.org/10.1177/0022185613489438

Lewer J (2015) *Not charted on ordinary maps: The Newcastle Steelworks closure.* North Melbourne: Australian Scholarly Publishing.

Luetjens J, Mintrom M and t Hart P (2019) *Successful public policy: Lessons from Australia and New Zealand.* Canberra: ANU Press. p. 550.

Martin R (2012) Regional economic resilience, hysteresis and recessionary shocks. *Journal of Economic Geography* 12(1): 1–32.

Maynard J (2004) *Awabakal word finder and dreaming stories companion*. Southport: Keeaira Press.

Moretti E (2012) *The new geography of jobs*. New York, NY: Houghton Mifflin Harcourt.

Net Zero Economy Agency (2024) *Impact analysis – Support for workers during the net zero transition*. https://oia.pmc.gov.au/sites/default/files/posts/2024/03/Impact%20Analysis.pdf

NSW Government (2018) *Hunter Regional Economic Development Strategy 2018–2022*.

NSW Government (2020a) *NSW electricity infrastructure roadmap, November 2020, building an energy superpower detailed report*. www.energy.nsw.gov.au/sites/default/files/2022-08/NSW%20Electricity%20Infrastructure%20Roadmap%20-%20Detailed%20Report.pdf

NSW Government (2020b) *Strategic statement on coal mining and exploration in NSW*. www.resourcesregulator.nsw.gov.au/sites/default/files/2022-11/strategic-statement-on-coal-exploration-and-mining-in-nsw.pdf

NSW Government (2021) *2021-22 NSW intergenerational report*. www.treasury.nsw.gov.au/sites/default/files/2021-06/2021-22_nsw_intergenerational_report.pdf

NSW Government (2024) *Future Jobs and Investment Authorities*. www.nsw.gov.au/regional-nsw/future-jobs-and-investment-authorities

NSW Government (n.d.) *Royalties for rejuvenation fund*. www.nsw.gov.au/regional-nsw/programs-and-grants/royalties-for-rejuvenation-fund

Orica (2023) *Orica Kooragang Island cuts greenhouse gas emissions by nearly 50 per cent*. www.orica.com/locations/australia-pacific-and-asia/australia/kooragang-island/news-media/orica-kooragang-island-cuts-greenhouse-gas-emissions-by-nearly-50-per-cent

Parliament of NSW (2019) *Sustainability of energy supply and resources in NSW*. Legislative Assembly Committee on Environment and Planning. www.parliament.nsw.gov.au/committees/inquiries/Pages/inquiry-details.aspx?pk=2542#tab-termsofreference

Parliament of NSW (2024) *Inquiry into beneficial and productive post-mining land use*. Legislative Council Standing Committee on State Development. www.parliament.nsw.gov.au/committees/inquiries/Pages/inquiry-details.aspx?pk=3046#tab-termsofreference

Phelan L and Crofts K (2022) 3 local solutions to replace coal jobs and ensure a just transition for mining communities. *The Conversation*, 17 January. https://theconversation.com/3-local-solutions-to-replace-coal-jobs-and-ensure-a-just-transition-for-mining-communities-174883.

Prime Minister of Australia (2024a) *Investing in a future made in Australia*. Media Release, 14 May. www.pm.gov.au/media/investing-future-made-australia

Prime Minister of Australia (2024b) *Solar Sunshot for our regions*. Media Release, 28 March. www.pm.gov.au/media/solar-sunshot-our-regions

Productivity Commission (2017) Transitioning regional economies, study report. Canberra. www.pc.gov.au/inquiries/completed/transitioning-regions/report/transitioning-regions-report.pdf

Pugalis L and Gray N (2016) New regional development paradigms: An exposition of place-based modalities. *Australasian Journal of Regional Studies* 22(1): 181–203.

Pulsford I, Howling G, Dunn R and Crane R (2013) The great eastern ranges initiative: A continental scale lifeline connecting people and nature. In: *Linking Australia's landscapes: Lessons and opportunities from large-scale conservation networks*. Canberra: CSIRO Publishing, pp. 123–134.

REMPLAN (2023) *Hunter Region Dataset*, May 2023.

Rey-Lescure O (2024) *Mining areas and power stations in and near the Hunter River Catchment*. Newcastle, Australia: University of Newcastle.

Rodríguez-Pose A (2018) The revenge of the places that don't matter (and what to do about it). *Cambridge Journal of Regions, Economy and Society* 11(1): 189–209.

Sheldon P, Junankar R and De Rosa Pontello A (2018) *The Ruhr or Appalachia? Deciding the future of Australia's coal power workers and communities*. Sydney: Industrial Relations Research Centre, UNSW Business School for CFMMEU Mining and Energy. https://me.cfmeu.org.au/policy-research/ruhr-or-appalachia-deciding-future-australias-coal-power-workers-and-communities (accessed 2 November 2021).

Tranter B and Foxwell-Norton K (2021) Only in Queensland? Coal mines and voting in the 2019 Australian federal election. *Environmental Sociology* 7(1): 90–101.

Turner JW (1982) *Coal mining in Newcastle, 1801–1900*. Newcastle: Council of the City of Newcastle.

Unger CJ, Everingham JA and Bond CJ (2020). Transition or transformation: Shifting priorities and stakeholders in Australian mined land rehabilitation and closure. *Australasian Journal of Environmental Management* 27(1): 84–113.

Weller SA (2017) The geographical political economy of regional transformation in the Latrobe Valley. *The Australasian Journal of Regional Studies* 23(3): 382–399.

Weller S, Beer A and Porter J (2024) Place-based just transition: Domains, components and costs. *Contemporary Social Science* 19(1–3): 355–374.

Weller S and Tierney J (2018) Evidence in the networked governance of regional decarbonisation: A critical appraisal. *Australian Journal of Public Administration* 77(2): 280–293.

Wells HC (1950) *The Earth cries out.* Sydney: Angus and Robertson.

Wiseman J, Campbell S and Green F (2017) *Prospects for a "just transition" away from coal-fired power generation in Australia: Learning from the closure of the Hazelwood Power Station* (CCEP Working Paper 1708).

6 From carbon capital to renewable energy superpower

Transforming the industrial hub of Gladstone, Central Queensland[1]

Amanda Cahill

Introduction

I watched Prime Minister Albanese announce yet more energy transition funding as part of his 'Future Made in Australia' plan in Gladstone, Queensland, with mixed emotions (Gladstone Today, 2024). Known for many years as Australia's 'Carbon Capital', multi-million dollar announcements of support for industries in Gladstone to decarbonise had become so regular in 2023 and 2024 that it was almost anti-climactic. Yet the Prime Minister's new announcement was also a remarkable symbol of how much the region had changed in only six short years.

In December 2018, barely a dozen people had attended a public forum organised by The Next Economy (TNE) on the future of energy in Gladstone, and in April 2021 the topic of transition was so contentious that the CEO of a large, state-owned energy company resigned after publicly acknowledging plans to shift to renewable energy and operate its coal plants 'much more flexibly, in response to market requirements'.

How had things changed so much in such a short period of time? How had some of the same people who in 2018 saw no need for change come to accept and even champion economic transformation less than six years later? This chapter addresses these questions, drawing on the experiences of TNE in working with different stakeholders to engage the community in energy transition conversations and planning processes between 2018 and 2024.

DOI: 10.4324/9781003585343-6

Gladstone: the Carbon Capital of Australia

The industrial town of Gladstone is built on the lands of the Bailai, Gurang, Gooreng-Gooreng, and Taribelang Bunda peoples at the Southern end of the Great Barrier Reef in Central Queensland (see Figure 6.1). For over 100 years after colonisation in 1847, Gladstone was a small town supported by cattle farming, fishing and coal exports. This changed in the 1960s when the Queensland Government established an alumina refinery. The investment in Queensland Alumina Limited stimulated a host of other developments, including the expansion of the Port of Gladstone, the construction of four coal-fired power stations and a wide range of carbon intensive industries including alumina and aluminium, cement products, ammonia nitrate and other chemicals.

During the 2010s, Gladstone's industrial base and port led to its emergence as a key site in the expansion of Queensland's gas industry, with three massive liquified natural gas (LNG) plants being developed. By 2023, the region was home to one of the biggest gas export terminals in the world, with 23 million tonnes of LNG shipped that year (LNG Prime, 2024). The same year, 67 million tonnes of coal was exported through the port, making it the world's fourth largest

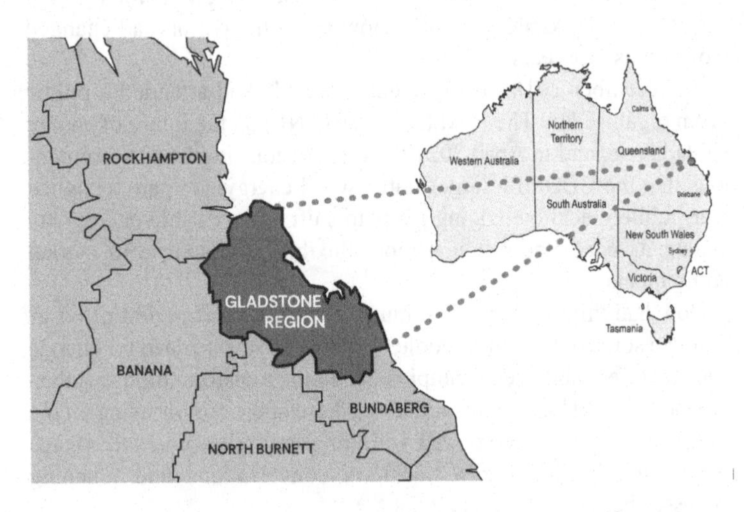

Figure 6.1 Map of the Gladstone Region.

coal-exporting terminal. Seventy per cent of this was metallurgical coal, bound mostly for steel mills in Asia (The Coal Hub, 2024).

Initiating the energy transition conversation (2018)

Attempts to engage the community in conversations about climate change and the energy transition started much later in Gladstone than in the other regions in this book. This was mainly because the coal-fired power plants in Central Queensland had been built later and their expected operating lives extended well into the 2030s and 2040s.

The first attempt was Gladstone Conservation Council's 'Repower Gladstone' campaign in 2018. Its focus was on the need for climate action and renewable energy, but it had limited traction. As a result, campaign partner the Australian Conservation Foundation (ACF) invited TNE to host an industry roundtable discussion and public forum in December 2018, focusing not on climate change but rather on the economic implications of a decarbonising global economy. But these two events were little more successful than Repower Gladstone: a combined total of just 18 people came to hear speakers from Beyond Zero Emissions (BZE) on the technical aspects of the energy transition, Voices of the Valley on the impacts of the Hazelwood Power Station closure (see Chapter 3) and the ACF on the impacts of climate change on the economy.

The small number of industry participants in attendance were polite but sceptical of the need for change, arguing that 'the coal industry would continue for at least another 50 to 100 years' and that 'renewable energy would never be able to power industry'. The few who called for action on climate change grounds in the public forum were met with disapproving looks from other community members.

In 2018, energy transition and climate action were simply not a priority for most people living across the region. The turning point came after a very public confrontation over the establishment of a new coal mine hundreds of kilometres north-west of Gladstone in April 2019.

Rising tensions (2019–2020)

Central Queensland had been a key battleground in Australia's 'climate wars' (Chapter 1) ever since the Indian conglomerate Adani announced plans to develop a massive new coal mine in the undeveloped Galilee Basin in 2010. Opposition to Adani's Carmichael Coal

Mine was coordinated through the 'Stop Adani' campaign, which worked for many years to build community opposition to the mine and associated rail line (Ritter, 2018).

Tensions came to a head in a violent clash in the small mining town of Clermont between environmental activists protesting against the Carmichael Mine and those wanting the mine to go ahead in the weeks leading up to the April 2019 Federal Election (Burt et al., 2019). This moment proved an important turning point, with many Central Queensland residents reflecting later that they felt pressured to 'pick a side' and either support the coal industry (and by extension, the local economy), or support the outsiders promoting climate action. The Deputy Prime Minister Barnaby Joyce characterised the mood at the time as: 'It's the people of the real world versus the latte-sipping, inner city trendies' (Clark, 2019). When the Coalition Government prevailed unexpectedly at the election, the 'Stop Adani' Convoy and the conflict in Clermont was frequently cited as a key turning point (Horn, 2019).

In Central Queensland, the pressure to choose a side had a chilling effect on any public discussion about the future of energy across the region. The State Government was concerned about growing social polarisation across the state on the future of coal, so funded TNE to convene an 'Energy Transition Roadshow' involving eight workshops across four regions, including Gladstone. The objective was to facilitate broader community understanding that the energy system was changing. In the workshops, energy experts[2] explained how the energy system was changing and how emissions could be reduced across all economic sectors, and representatives from the Latrobe Valley and Hunter Valley explained how early power station closures and uncertainty about the future of coal was affecting them.[3]

In March 2020, over 60 people attended the workshops in Gladstone. Participant numbers were still low and tensions were high, but the tone of the conversation had noticeably shifted since 2018. Some community members who had previously been sceptical about climate change asked questions about the link between fossil fuels and climate change. Others asked whether the catastrophic 2019 Black Summer Bushfires were linked to a changing climate. The mix of participants had also changed. This time most of the attendees were drawn from heavy industry, energy companies, or different levels of government.

The sense that the region needed to prepare for change grew during 2020, with unexpected announcements from key trading partners Japan and South Korea that they were aiming for net zero economies by 2050. The impact of these announcements were reinforced by a growing number of reports by leading climate and environmental groups focused on the economic and employment opportunities for industrial areas willing to transition to renewable energy (e.g. BZE, 2020; Cahill, 2020; CEC, 2020).

TNE started to receive requests from industry, government and civil society groups for support to explore the regional implications of the energy transition. Behind closed doors, boards of major energy companies expressed concerns about the lack of constructive public conversations about the energy transition, as well as risks to their regional operations if they could not find ways to reduce emissions and switch to renewable energy. Fossil fuel companies reported that they were under increasing pressure from insurers and financiers to demonstrate how they would reduce emissions. They also noted increasing risks that political turmoil surrounding the national climate change conversation would mean that regions like Gladstone would miss out on new investment needed to move to renewable energy.

Energy executives were not alone in these concerns. State Government officials, concerned about the urgent need for a comprehensive energy plan for Queensland, were becoming increasingly frustrated with the divisive politics stifling public conversations about the future of energy. Similarly, environment and union movement leaders worried that future climate campaigns could spark another round of conflict in Central Queensland given the ongoing tensions across the region.

Breaking the silence: the Central Queensland Energy Futures Summit (2021)

In response to these concerns, TNE proposed holding a summit in Gladstone, inviting representatives of all stakeholder groups to present their ideas on the actions needed to prepare the region for change.

Surprisingly, all groups embraced the idea. An unlikely alliance of Stanwell Energy (Australia's third largest emitter), CleanCo (a new, publicly owned renewable energy developer), the Clean Energy Finance Corporation (Australia's 'green bank'), the ACF (Australia's largest conservation group) and CQ University joined forces to sponsor the event. The political risk lay with TNE, perceived by many to be more neutral because it was an external organisation focused on transition as an economic (rather than a climate) challenge.

The fact that such disparate groups were willing to sponsor a summit on the future of energy in Central Queensland generated intense local, national and international interest. Participation was limited to 150 people due to COVID restrictions and the event was over-subscribed. Attendees included representatives of energy companies (fossil fuel and renewable), heavy industries, all levels of government, training providers, unions, environment groups and Traditional Owners. To facilitate a more open discussion, politicians were not invited.

Held over two days, the Central Queensland Energy Futures Summit (Whittlesea, 2021) was structured to support participants to deepen their understanding and work together to analyse how the region could manage change to:

- Ensure a stable, renewable energy supply;
- Support workers impacted by the shift from fossil fuels to renewable energy;
- Diversify the economic base;
- Build a sustainable, renewable hydrogen industry; and
- Ensure that the community benefited from any new developments.

Facilitators from TNE interviewed many of the attendees in advance of the summit, and most expressed significant concerns. Some were worried they would lose their job if they attended. Others, who had been burnt by previous interactions, swore that they would never be 'in the same room' with each other. Given the level of concern amongst participants, the atmosphere in the auditorium at the start of the Summit could only be described as tense. As one participant later reflected: 'You really could cut the tension with a knife. Everyone was nervous about what would happen next'.

The tension in the room only started to ease after Richard van Breda, the CEO of the Stanwell Energy, explained why they had decided to sponsor the event:

> Australia is undergoing a major energy transition and it's happening at a rapid pace. The energy market is shifting from fossil fuel generation to renewable energy and storage… We will operate our

> coal-fired power stations much more flexibly, in response to market requirements. This may include seasonal storage of our generating units, or placing units into standby mode so they can be quickly returned if the market needs them... we need to make changes in order to remain relevant to our customers.

That the CEO of Australia's third largest emitter Stanwell would dare to publicly acknowledge the need to prepare for the phase out of coal and the introduction of renewable energy in Central Queensland signalled a turning point for the rest of the participants. As one industry representative later reflected: 'When Richard spoke, it gave us permission to get on with it'.

And get on with it they did. To the surprise of most, participants were able to work together despite differing opinions by focusing on the long-term benefit of the region. Participants expressed gratitude for the content and also the opportunity to speak and work collaboratively with people they did not normally associate with.

> It was a great start... Everyone was willing to listen to everyone's point of view. Thank you for Next Economy for having the courage to put the event on!

Media coverage of the event was also overwhelmingly positive, with most journalists commenting on how remarkable it was that such an event was being held in Central Queensland at all given the level of tension around the topic of energy and climate change. A regional radio host commented after an interview on the Summit:

> Are you really telling me that it's going well? With all those people together talking about energy? That's gotta [sic] be a first.

Given the apparent success of the event and positive media coverage, it came as a huge shock to everyone when it was announced that Richard van Breda resigned only two days later, after refusing to bow to political pressure to retract his statement (Smee, 2021). Media attention surrounding the event reached new heights, with journalists seizing on the resignation as yet another example of Australia's broken politics around energy and climate and the risk to those who were 'too honest about the energy transition from coal-fired power to renewable energy' (Boyd, 2021).

To make matters worse, Callide C power station in the nearby town of Biloela exploded only a month after the Summit, leading to state-wide power shedding and temporary blackouts. The prospect of a renewed conversation about energy transition seemed decidedly remote.

But Richard's resignation and the blackouts after the Callide C incident seemed to have the opposite effect, with local leaders talking more publicly about the need for the region to transition. Within months, Rio Tinto announced plans to decarbonise its aluminium operations. The Gladstone Ports Corporation started talking about plans to become less reliant on fossil fuel exports. Other Gladstone-based companies including Cement Australia and Orica started to develop decarbonisation strategies. The Gladstone Regional Council launched an engagement process to develop a ten-year Energy Transition Roadmap and BZE started to work with industry leaders to identify actions needed to establish a Renewable Energy Industrial Precinct (BZE, 2022). The ACF joined with the Queensland Community Alliance (a coalition of union, environment and faith-based groups) to take the transition conversation out into the broader community as part of the 'Real Deal' project (QCA, 2023).

The momentum was helped by a surge in interest from international investors who had read media coverage of the Summit.[4] Local confidence was also boosted by a suite of new renewable energy developments and the announcement by Future Fortescue Industries that it would build a new electrolyser plant in the region to service the emerging green hydrogen industry.

Taking the reins: Gladstone Region Economic Transition Roadmap (2021–2022)

Following the Summit, Gladstone Regional Council began an 18-month-long participatory planning process in partnership with TNE to develop a ten-year Gladstone Region Economic Transition Roadmap (TNE, 2022).

Council had adopted the enabling resolution a year earlier in June 2020 but not all Councillors supported it, and as the end of 2020 approached no action had been taken. In response, Council's economic development officer Garry Scanlan approached TNE to run a workshop with Council to explore what would be involved in developing a transition plan.

All nine councillors and key managers assembled for the workshop in December 2020. Around half the Councillors remained unconvinced that a transition plan was a good idea. One summed up the mood at the outset of the workshop:

> Can't you just do what other consultants do and do a few interviews and then go away and write the plan yourself? Then we can say that we've done it if anyone asks, without having to cause any more problems in the community.

Perspectives amongst Councillors started to shift after a presentation by Tim Buckley, a financial analyst from the Institute for Energy Economics and Financial Analysis. Buckley's presentation showed that international investment was moving away from coal into new renewable energy industries such as Green Hydrogen. Even those who supported the coal industry conceded that maybe they should develop a transition plan 'just in case' change was really coming.

Despite this renewed commitment, it wasn't until after the Energy Futures Summit four months later that Gladstone Regional Council gave TNE permission to start the process. One of the Councillors who had been reluctant to undertake the transition planning process later reflected:

> I thought coal would be around for decades. But at [the Summit], I remember the moment I suddenly got it. The world was changing. We might not like it, but we needed to figure out what we were going to do. Because we have all these industries that need coal, so how are we going to power them with something else?

TNE worked closely with the Council to co-design a participatory planning process involving iterative cycles of co-design, engagement, reflection and action. By the end of the process, TNE had engaged with 264 people representing First Nations groups, industry, government, workers, environment groups, social services and the general community across 13 separate engagement activities.

The Energy Transition Roadmap (TNE, 2022) was launched in November 2022. The most significant outcome was the high level of local commitment to translating it into action. Even before the launch, Council officers had met with key industry leaders to discuss how they could work together to implement the roadmap and presented on the roadmap approach and priorities to five other councils across the region to initiate a conversation on joint advocacy priorities.

The positive and public steps taken by the Gladstone Regional Council attracted further interest from international investors wanting to establish renewable industries in the area. Examples included proposals to develop renewable hydrogen, batteries and a range of green chemicals. A senior manager at the Council later remarked:

> With two or three trade delegations coming from all over the place every week, even from Taiwan, Japan and other places, we started to joke that Council should set up its own tourism business.

From Gladstone to the world (2022–2024)

By 2022, Gladstone was centre stage in conversations about how Australia could achieve net zero emissions, and since the election of the Federal Labor Government, has become a regular site for state and federal government energy transition consultation processes and announcements.

Inspired by Gladstone's Energy Transition Roadmap, other industrial and coal regions around Australia started approaching Gladstone Regional Council for support to develop their own transition plans. The Council also initiated a joint campaign with other local governments to advocate for policy and funding measures to support regions to plan for and manage the transition to net zero. This included contributing to advocacy efforts calling for a national body to support regions to manage the transition—the Net Zero Economy Authority.

Gladstone's influence even spread internationally, with the Mayor, Council staff and TNE invited to present on Gladstone's transition plan at government and industry meetings in Taiwan, Germany, the Netherlands and the United States.

A range of companies operating across Gladstone embraced the recommendations outlined in the roadmap, engaging external consultants including Arup, Climate-KIC and Climateworks to undertake detailed economic and investment modelling. One chemicals manufacturer reflected: 'The roadmap showed us that the community was ready for us to take action'.

Community organising efforts by ACF and the Queensland Community Alliance also ramped up in 2023 with the launch of a report detailing regional community concerns and aspirations. The Alliance also negotiated a regular roundtable discussion with Council, in part to support the implementation of the Energy Transition Roadmap in a way that benefitted the broader community.

While the Energy Transition Roadmap and other engagement activities have helped to strengthen the capacity and confidence of leaders across different stakeholder groups to accept and manage the transition, many challenges remain.

Many people across the community have still not been directly engaged in the transition conversation, as most of the engagement activities targeted leaders of stakeholder groups directly affected by the energy transition. Some community members have expressed concern that transition could add to cost of living pressures, housing shortages, substance misuse and other social and economic challenges, referencing the negative impacts that many people experienced during the LNG expansion (QCA, 2023).

Others remain sceptical that the transition will happen at all. Despite frequent announcements of new developments and construction projects, long approval and development timelines have left some questioning whether the transition is 'just talk'. Coal companies continue to generate record profits, and many leaders who are publicly supportive of the roadmap and renewable future (including some Councillors) continue to insist that there will be a global demand coal and gas for decades to come. Most recently, the Federal MP Colin Boyce organised anti-wind forums and sought to build support for the construction of a new nuclear plant in the region, sowing further uncertainty and confusion about the energy transition (Boyce, 2024).

Yet despite these challenges, local leaders continue to promote Gladstone's net zero vision. Mayor Matt Burnett summed it up in a 2023 promotional video produced by the Council (GRC TV):

> In the Gladstone Region we are leading the way in planning for a new economic future. A future that will help drive Australia to achieving net zero emissions… This means a future where renewable energy shapes every aspect of our region: our economy, our industries, our work opportunities, and most importantly, the quality of life our communities can offer.

Championing change—the role of local leadership

Change is rarely a neat, linear process. It is more often the outcome of a complex interplay of multiple events—both planned and unexpected. National climate campaigns, ambitious emissions targets amongst Australia's trading partners and unexpected external shocks such as the COVID pandemic and 2019 Black Summer Bushfires all contributed to a growing awareness and acceptance that the world was changing, and that Gladstone would have to change with it. What stands out as unique to the Gladstone experience is the pivotal role played by leaders in local government and industry in bringing the community together to host and shape the conversation about transition.

From Gladstone Regional Council adopting a resolution to develop a transition plan in 2020 to heavily coal-invested Stanwell Energy sponsoring the Energy Futures Summit; from companies announcing their plans to decarbonise before the Australian Government had even committed to a net zero target, to later efforts by local government and industry leaders to advocate for policies at a state and federal level to support decarbonisation, it was leadership from 'champions of change' in formal positions of power that made the difference between paralysis and action to secure a better future for the region.

While in other parts of Australia the energy transition was put on the agenda and carried by environment groups, informal community interest groups and/or unions over years, in Gladstone the conversation only really started once leaders in formal organisations came together to discuss the economic implications of change. This was driven by context. By the time the transition conversation was starting in Gladstone, the battlelines in Australia's 'Climate Wars' had long been drawn and local residents were ready to believe the media narrative that environmental groups were outsiders who did not have the interests of the region at heart. As expressed by local Senator Matt Canavan in 2019:

> …it's incredibly patronising because these out-of-town people think they have the right to come into someone else's community and tell them they're all evil and wrong and need to be sacked or re-skilled.
>
> (Crowe, 2019)

Unions had also been largely sidelined in the public conversation about transition in the region, wary of undermining the interests of local coal workers. For the community to overcome this entrenched polarisation, unlikely local leaders needed to step forward to champion the need to prepare for transition. These leaders needed support, which they found through engagement approaches which:

1. were designed to meet people where they were at to support them to manage change, rather than push for change;
2. the leaders themselves were involved in designing; and
3. situated what was happening locally within national and international transition conversations.

These three principles were the key lessons learned through Gladstone's transition to date and are expanded further below.

Meeting people where they are at

The focus of TNE's engagement activities started with the issue people across the community were most concerned about: the potential economic impacts of the rest of the world moving to net zero emissions. The starting point for discussions was that change was inevitable and decarbonisation efforts were already impacting the region. This was a different message to that of most climate and environment groups, which aimed to mobilise as many people as possible in order to build pressure on decision makers to accelerate climate action.

Starting with what people cared about most was inspired by the Stages of Change model developed by Prochaska and DiClemente (1983) after studying behaviour patterns amongst people attempting to quit tobacco smoking (see Figure 6.2). This model suggests that people will start to consider changing their behaviour when it impacts on something they personally care about. The model also suggests that people at different stages of awareness and acceptance of the need to change require different communication techniques and support services. For example, people who are in the 'pre-contemplative' stage (i.e. not even contemplating giving up smoking) need different strategies to those who are in the 'contemplative' and 'preparation' stages of wanting to quit, or for those that have already been through the 'action stage' and so want to maintain a healthier lifestyle.

TNE designed engagement approaches based on the stage the community was at, exploring both how they would manage the risks they faced and the potential opportunities if they were able to manage change well. For example, between 2018 and 2020, when both awareness and acceptance of a fossil fuel transition were low, TNE organised public

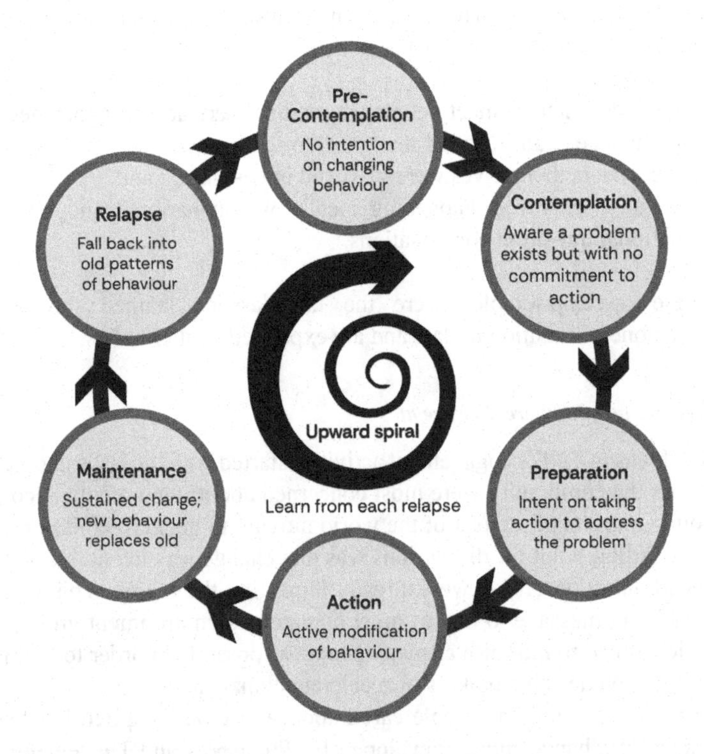

Figure 6.2 The 'Stages of Change' model.

Source: Adapted from https://socialworktech.com/2012/01/09/stages-of-change-prochaska-diclemente/

presentations by respected experts which showed how global efforts to decarbonise the energy sector were already starting to influence capital flows away from fossil fuels and towards renewable energy. These presentations also highlighted the new economic opportunities for regions able to make the transition to renewable energy.

In early 2020, when it was difficult to even mention the energy transition in public because it was being framed as an environmental issue pushed by 'inner-city latte sippers', TNE hosted speakers from the Hunter and Latrobe Valleys to show that change had already come to other industrial regions. Hearing how other regions were being impacted by changes in the energy sector and what they were doing to respond made it 'real' for Gladstone leaders and showed them that they too could have agency rather than just wait for change to happen to them.

During the Energy Futures Summit in 2021, when most people were already starting to take small steps, TNE arranged for local leaders to present on actions they were already taking to manage change. Positive interactions at the Summit and media coverage broke through the silence holding back decision making by celebrating what was already happening. This, in turn, bolstered a stronger sense of agency, pride and possibility across the region.

While developing the Energy Transition Roadmap between 2021 and 2022, the approach shifted again to enable a more detailed and technical discussion about actions required to manage change across all sectors. Once the roadmap was launched, this involved TNE taking a step back from direct involvement to enable local leaders to start to advocate for what they needed to manage change.

By meeting people where they were at over a number of years, TNE created space for people to reflect on what they really wanted for their community. Supporting people to explore the risks and opportunities of change as well as their own strengths built their identity as a strong regional industrial hub with the opportunity to be part of the solution to climate change. When they understood what people in their own community were already doing to manage change, Gladstone's regional sense of identity shifted over time from 'Australia's Carbon Capital' to an 'industrial ecosystem with a long-term vision' that can 'help decarbonise the world' (GRC TV, 2023).

This is not to say that the trajectory in Gladstone was trouble-free. As reflected in the Stages of Change model, change is rarely a linear process, particularly when people do not have access to ongoing support. Recent reports of rapidly growing scepticism across the region about the energy transition are testament to the need for ongoing engagement processes to support change (QCA, 2023). Efforts to organise, update, communicate, and connect with decision making at different levels need to be continuous. Fortunately for Gladstone, the Council, industry and the Community Alliance continue to engage with the broader community. But it is yet to be seen whether this local commitment will be able to withstand growing opposition to renewable energy.

Deepening opportunities for participation

The second factor that supported community leaders was the participatory nature of the engagement activities. Long accepted by community development practitioners as key to achieving sustainable development outcomes, participatory approaches encourage local people to analyse their own situation and develop ideas to address their own challenges.

TNE adapted the Participatory Action Research (PAR) approach[5] to support key stakeholders to not only participate in interactive engagement exercises but also in the design and delivery of engagement activities and subsequent actions. This meant that each intervention over the six-year period from 2018 was co-designed with, and sometimes even facilitated by, local stakeholders. While the structure of different activities was quite similar, the content, presenters and questions explored were tailored according to feedback from stakeholders.

While the goal was to maximise meaningful participation, not everyone was engaged in every activity. Instead, engagement activities at each stage were designed with the principle of 'right people, at the right time, in the right way'. Wherever possible and appropriate, those closest to the energy transition were targeted so that they could understand and provide meaningful input into decision making processes.

As the capacity, knowledge and confidence of different stakeholders grew over time, local leaders took on more responsibility for leading conversations: first with the community and later with state and federal policymakers. This period, after the public launch of the Energy Transition Roadmap, was where TNE's role shifted from design and facilitation to advice and relationship brokering to support local leaders to implement their plans and advocate for the community.

TNE's not-for-profit status was key in alleviating the typical pressures most consultants face, because it meant it could draw on philanthropic support to cover the costs of the additional time and engagement work. This was essential to allow for the careful design and implementation of a genuinely participatory planning process, which remain rare in Australia. It enabled TNE to commit to working with the community over the long term, including responding to requests for support regardless of whether there was a paid contract in place. This flexibility meant that TNE could reduce some of the practical barriers to participation, for example by scheduling the timing of events to when people were most likely available and providing a small stipend to some First Nations participants who had travelled long distances to attend.

The decision to focus early engagement activities on local leaders in government, industry and other formal institutions meant that fewer people were engaged than in typical public engagement or community organising activities. This has meant the process has not led to the creation of a broad based, formal institution to oversee the implementation of energy transition activities. Many stakeholders in Gladstone have advocated for the establishment of a regional transition authority or committee to oversee the implementation and coordination of transition related initiatives using a model similar to that employed by the Spanish Government to support coal regions (ITJ, 2022). While local leaders are doing what they can, without an institution that has the authority to bring all stakeholders together to regularly review plans and progress, many are now reporting that there is an increasing lack of coordination as the pace of change accelerates.

From the local to the global: the role of external catalysts

While the participatory approach supported local leaders to shape the energy transition conversation, they did not act alone. A range of external actors and national and global trends contributed to the shift in community attitudes and their willingness to confront and manage change.

In Gladstone, the first external influence that proved a powerful catalyst for change was the 2019 clash between protestors and supporters of the Adani Carmichael mine in Clermont. Though the initial aftermath was a strong outpouring of support for regional jobs (and the fossil fuel industries many were associated with), this clash also forced local leaders to ask hard questions about the economic risks that decarbonisation might pose.

Timely inputs from a number of NGOs and think tanks around this time on the technical, economic and financial aspects of transition attracted the attention of business and government leaders. TNE was able to draw on this work by hosting a variety of speakers, creating space for people in Gladstone to think differently about the challenges they faced and start analysing what it would take to decarbonise different sectors of the economy while supporting workers and the community.

Notably, the most influential speakers were often community members from the Latrobe Valley and Hunter Valley, who shared their experience of the impacts of early closures and an uncertain future for coal mining. Not formally recognised 'experts', they brought insights into setting up local alliances and engaging communities to plan and manage change. Hearing from 'real people' going through a similar experience 'made transition real' and showed it was possible for local leaders to support the Gladstone community to navigate their way through it.

International investors were also highly influential in providing the impetus for change. Given the region's economic dependence on fossil fuel exports, local leaders took note when major importers Japan and Korea announced net zero targets in 2020 and when the European Union announced plans to develop a Carbon Border Adjustment Mechanism. While these announcements alerted people up to the threat of change, the approaches from international investors after the Energy Futures Summit in 2021 and launch of the Energy Transition Roadmap in 2022 highlighted the range of new economic opportunities, boosting the confidence of local leaders to be more proactive and vocal in support of transition.

National and international media coverage also played an important role in shaping the transition conversation across the region and beyond. While both local leaders and TNE were initially wary of journalists wanting to cover events, over time it became clear that the intense media interest was significant both locally and more broadly. It showed people in Gladstone that what they did mattered, and that the world was paying attention. The overwhelmingly positive coverage of key events encouraged local leaders to speak publicly about their plans, and the sense that 'other people were watching' encouraged accountability.

The national media coverage featuring the voices of local Gladstone leaders talking about the need for support for all coal regions in the lead up to the 2022 Federal Election played an important role in undermining concerted attempts by representatives by the Liberal and National Parties to argue against the adoption of a net zero target and policies to support decarbonisation. As a leading climate campaigner later reflected:

> TNE's work in Central Queensland undermined efforts by politicians such as Barnaby Joyce and Matt Canavan to weaponise the idea of transition and equate 'transition' with 'unemployment'. Projects like the one in Gladstone have been fundamental to

> shifting the national debate around renewables and fossil fuels. No longer do we have communities as a whole believing the denialism that coal and other fossil fuels will be with us forever. No longer do we have a small group of political leaders who do not believe in climate science being able to raise these fears to any credible level in national and state politics.

Similarly, a high-level ministerial advisor in the Labor Party later reflected that the level of positive media coverage about Gladstone in the lead up to the 2022 Federal Election gave the newly elected government the 'mandate it needed to go hard and fast to accelerate the transition'.

The positive impact of open, public debate about the energy transition that connects particular regions to national and global trends is not limited to the Gladstone experience. Open, transparent and public dialogue has been demonstrated to be an effective antidote to polarisation and stimulus for accelerated action elsewhere around the world, including Germany's 'Coal Commission' dialogue process (Agora Energiewende, 2019), the Climate Assemblies in Barcelona and other parts of Europe (Platoniq, 2024).

Conclusion

Locally led, participatory approaches have been central to community development theory and practice for decades, yet genuinely participatory approaches remain rare in Australia. Despite the lip service many consultants and governments pay to the importance of community participation, most engagement processes remain at the 'manipulation' or 'informing' end of Arnstein's (1969) widely cited participation ladder, rather than the more empowered 'citizen control' approaches that expand people's agency to challenge and transform entrenched power relations. In large part, this is because facilitating meaningful, open and inclusive participation requires significant time and resources. Yet when done well, this investment both sets communities up to successfully navigate change *and* benefits governments and industry actors which might otherwise be threatened by a community determining their own goals.

It remains to be seen whether the current Australian Government will learn from the experience of Gladstone and provide adequate resources and support to enable regional leaders to plan and manage change. The final report on a recent senate inquiry into the role of the newly established Net Zero Economy Agency (NZEA; Parliament of Australia, 2024) emphasises that its primary mandate is to support *workers* when coal-fired electricity plants close, and fund heavy industry to decarbonise their operations and stimulate new investment in green manufacturing. At the time of writing, there are no dedicated government funding programmes supporting regions to develop their own transition plans or establish local committees to coordinate the transition. Furthermore, the only government funded regional transition authority—the Latrobe Valley Authority—has been defunded and will cease to exist from October 2024.

Given the benefits demonstrated by participatory planning activities in Gladstone, this lack of targeted Federal Government support is perplexing. Perhaps it is because Gladstone's efforts were funded by a combination of philanthropy, industry sponsorship and small amounts of state government funding, and so the processes remain mostly invisible. Or, even more concerningly, perhaps, as has been suggested in confidence by three senior public servants, the lack of support for regions to lead planning and coordination activities is that the Federal government 'doesn't want to lose control' or 'doesn't want to be put in a position where they can't deliver what the community wants' particularly now that there is 'fast growing opposition to renewables'.

The irony is that Gladstone, the region which arguably had the most targeted support to develop a transition plan in advance of any closures is actually one of the Federal government's biggest allies in supporting the development of renewable energy and green industries. The local government in particular has been concerted in its efforts to push back against attempts by the local Federal MP to whip up community opposition to wind farms and replace coal power plants with nuclear power (Evans, 2024).

Many challenges remain and there is more work to be done before Gladstone can fully seize the economic opportunities that a smooth and just transition could present, but leaders across the region have now demonstrated time and again that they are willing and able to respond to external events with confidence. They understand what their community wants, and now have the confidence and capacity to navigate the messy waters of transition in the hope of building a stronger, renewable economic future.

Notes

1 Amanda is the CEO of The Next Economy, an NGO contracted to facilitate engagement activities on the energy transition in Gladstone between 2018 and 2024. The insights presented in this chapter draw heavily on data and observations gathered through engagement activities she facilitated during this period with TNE staff Emma-Kate Rose, Emma Whittlesea, Sandi Middleton, Angela Heck, Jai Allison and Lisa Lumsden.

2 Speakers included Vanessa Petrie (Beyond Zero Emissions), Verity Morgan-Schmidt (Farmers for Climate Action) and Kristy Waters (Community Power Agency).

3 Between 2019 and 2021, a range of regional representatives participated in workshops and presentations, including Wendy Farmer (Voices of the Valley), Dan Musil (Earthworker Cooperative) and Karen Cain (Latrobe Valley Authority) from the Latrobe Valley, and Warrick Jordan (Hunter Jobs Alliance) and Joe James (Hunter Joint Organisation) from the Hunter Valley.

4 In the week following the Summit alone, Trade and Investment Queensland staff reported three new inquiries from Japanese and South Korean companies that had cited positive reports on the Summit as the reason for contact.

5 The genesis of PAR is contested. The approaches adopted by The Next Economy are largely influenced by the work of Paulo Freire (1972).

References

Agora Energiewende und Aurora Energy Research (2019) *The German Coal Commission. A roadmap for a just transition from coal to renewables.* www.agora-energiewende.de/fileadmin/Projekte/2019/Kohlekommission_Ergebnisse/168_Kohlekommission_EN.pdf

Arnstein S (1969) A ladder of citizen participation. *Journal of the American Institute of Planners* 35(4): 216–229.

Beyond Zero Emissions (2020) *Million jobs plan.* June 2020. www.bze.org.au/research/report/million-jobs-plan

Beyond Zero Emissions (2022) *Gladstone renewable energy industrial precinct briefing paper.* April 2022. www.bze.org.au/impact/reips/gladstone

Boyce C (2024) Boyce set to host energy forums. *Mirage News*, 10 April. www.miragenews.com/boyce-set-to-host-energy-forums-1211867/

Boyd T (2021) Mystery surrounds Stanwell CEO's exit. *Australian Financial Review*, 27 April. www.afr.com/chanticleer/mystery-surrounds-stanwell-ceo-s-exit-20210426-p57mj0

Burt J and McGhee R (2019) Horse rider arrested after clash with anti-Adani protesters that left woman injured in Clermont. *ABC News*, 28 April.

www.abc.net.au/news/2019-04-28/adani-protester-injured-in-clermont/11052940

Cahill A (2020) *What Queensland wants: Regional perspectives on building a stronger economy. The Next Economy*, August 2020. https://nexteconomy.com.au/work/what-queensland-wants-report/

Clark A (2019) Barnaby Joyce doubles down on climate change as an elitist issue. *Australian Financial Review*, 29 March. www.afr.com/politics/federal/barnaby-joyce-doubles-down-on-climate-change-as-an-elitist-issue-20190328-p518gi

Clean Energy Council (2020) *Clean energy at work.* Clean Energy Council. 2 June. https://assets.cleanenergycouncil.org.au/documents/resources/reports/Clean-Energy-at-Work/Clean-Energy-at-Work-The-Clean-Energy-Council.pdf

Crowe D (2019) Resources Minister backs new coal plant as Labor reconsiders climate policy. *Sydney Morning Herald*, 29 May. www.smh.com.au/politics/federal/resources-minister-backs-new-coal-plant-as-labor-reconsiders-climate-policy-20190528-p51s31.html

Evans J (2024) Nuclear proposal rejected by premiers, who say Dutton has no power to life state nuclear bans. *ABC News*, 19 June. www.abc.net.au/news/2024-06-19/premiers-reject-nuclear-proposal-nuclear-bans/103997020

Freire P (1972) *Pedagogy of the oppressed*. London: Penguin.

Gladstone Today (2024) PM Visit: Future made in Gladstone. *Gladstone Today*, 18 April. https://gladstonetoday.com.au/news/2024/04/18/pm-visit-a-future-made-in-gladstone/

GRC TV (2023) Gladstone region industrial ecosystem. Gladstone Regional Council. www.youtube.com/watch?v=t_ZhIZeLxo8

Horn A (2019) Why Queensland turned its back on Labor and helped Scott Morrison to victory. *ABC News*, 19 May. https://amp.abc.net.au/article/11122998

Instituto para la Transicion Justa (2022) Spain, towards a just energy transition. Executive Report. July 2022. www.transicionjusta.gob.es/Documents/Noticias/common/220707_Spain_JustTransition.pdf

LNG Prime (2024) Gladstone LNG exports slightly up in 2023. *LNG Prime*, 10 January. https://lngprime.com/australia-and-oceania/gladstone-lng-exports-slightly-up-in-2023/101864/

Parliament of Australia (2024) Net Zero Economy Authority Bill 2024 [Provisions] and the Net Zero Economy Authority (Transitional Provisions) Bill 2024 [Provisions] Report. May. www.aph.gov.au/Parliamentary_Business/Committees/Senate/Finance_and_Public_Administration/NetZeroBills2024/Report?>

Platoniq Foundation (2024) Towards a climate democracy: Seven citizens' assemblies and a manifesto. *Wilder Journal* #2. https://journal.platoniq.net/en/wilder-journal-2/futures/marea-deliberativa-manifiesto-asambleas/

Prochaska J and DiClemente C (1983) Stages and processes of self-change of smoking: Toward an integrative model of change. *Journal of Consulting and Clinical Psychology* 51(3): 390–395.

Queensland Community Alliance (2023) *A real deal for Gladstone: A listening report to shape economic and social transition*. Sydney Policy Lab. July 2023. www.sydney.edu.au/sydney-policy-lab/news-and-analysis/latest-news/2023/07/28/new-community-listening-report-a-real-deal-for-gladstone.html

Ritter D (2018) *The coal truth: The fight to stop Adani, defeat the big polluters and reclaim our democracy*. Crawley: UWA Publishing.

Smee B (2021) CEO quit Queensland's biggest power generator after energy minister complained to board. *The Guardian*, 28 April. www.theguardian.com/australia-news/2021/apr/28/ceo-quit-queenslands-biggest-power-generator-after-energy-minister-complained-to-board

The Coal Hub (2024) Australian coal relatively successful in finding new markets. *The Coal Hub*, 10 June. https://thecoalhub.com/australian-coal-exports.html

The Next Economy (2022) Gladstone region economic transition roadmap. October 2022. https://nexteconomy.com.au/work/gladstone-regions-economic-transition-10-year-roadmap/

Whittlesea E (2021) Central Queensland Energy Futures Summit Report. *The Next Economy*, May 2021. https://nexteconomy.com.au/work/central-queensland-energy-futures-summit/

7 Making regional energy transitions possible and making them just

Lessons from recent Australian experience

Gareth A.S. Edwards, John Wiseman and Amanda Cahill

Introduction

As this book goes to press in late 2024, energy transitions in Australia are well underway after a watershed federal election in May 2022. The idea that Australia stands to gain from playing a significant role in the new decarbonised economy has moved from marginal to mainstream and governments at multiple levels have become enthusiastic cheerleaders for transition policies.

Our aim in this book has been to explore how transitions were put on the agenda in the decade before 2022. The book has examined five key regions which dared to contemplate transition despite an extremely difficult political context. Our contributors have highlighted the successes of these initiatives, reflected critically on their failures and paused to consider the serendipitous connections which facilitated them. In short, this book has explored through these case studies how regional energy transition in Australia moved from impossible to possible.

But there are already indications that as the transition has begun to materialise and gather pace, justice has slipped out of focus. Disquiet and even open conflict has emerged not so much over the pace of fossil fuel closures but over the planning, approval and construction of renewable energy projects and the expansion of the electricity transmission lines needed to connect them to the grid. As governments and project proponents have rushed to bring new renewable energy

DOI: 10.4324/9781003585343-7

projects online, communities—especially those whose history has been bound up in providing fossil fuel energy—have been left asking what the sudden influx of green capital means for their economic and social fabric after the rush of development subsides.

If the notion of a 'just transition' was originally envisaged as a mechanism to overcome the 'jobs vs. environment' narrative that so often pitches the labour movement against the environment movement, the emerging challenge is of speed vs. justice. In the context of a progressing transition, the principle of 'leaving no workers and no communities behind' is not merely an ethical principle. It is also a practical imperative because justice is the basis for public support of energy transition plans and strategies, particularly in the regional communities most directly affected (Ciplet and Harrison, 2020).

Continued focus on justice is essential to counterbalance the temptation to assume that economic benefits will 'trickle down' throughout the community and that they will also provide social benefits (Weller et al., 2024). In fact, realising the economic and social benefits of the energy transition requires planning and engagement to identify and provide employment opportunities, infrastructure and social services to both communities transitioning away from fossil fuels and communities hosting the renewable energy and storage projects that are replacing them. These plans must ensure that livelihoods are maintained both during the transition process and for the long term.

The five case studies explored in Chapters 2–6 of this book have highlighted some key characteristics that this planning and engagement must (and *must not*) have. After briefly taking stock of how the transition discussion has developed in Australia since 2022, this concluding chapter distils out the key lessons from them. There are lessons for both researchers and policymakers. The former will be anxious to understand how the theory of 'just transition' rubs up against reality, while the latter will be concerned with what Australia can and should do as it enters a new phase of its transition journey and/or what other countries and regions grappling with the challenge of energy transition can learn from these Australian experiences.

Recent Australian energy transition trends and challenges

As highlighted in Chapter 1, ongoing reductions in the relative cost of renewable energy combined with growing support for stronger climate action have strengthened public support for a transition from fossil fuels to renewable energy in Australia. Chapters 2–6 showed that communities in fossil fuel-producing regions were at the leading edge of grappling with this impending energy transition.

The biggest shift occurred at the national level following the watershed May 2022 federal election, as the Albanese Labor government sought to implement a range of transition-related measures. One of its earliest actions was passing the *Climate Change Act (2022)* to codify into law national emissions reduction targets of 43% below 2005 levels by 2030 and net zero emissions by 2050. More recently, it passed the *Net Zero Economy Authority Act (2024)* which establishes a Net Zero Economy Authority to 'promote economic transformation as Australia transitions to a net zero emissions economy'.[1] Also in 2024, the Commonwealth budget was anchored with a $22.7 billion *Future Made in Australia* package to 'realise Australia's potential to become a renewable energy superpower, value-add to our resources and strengthen economic security by better attracting and enabling investment'.[2] This flagship package was accompanied by $96 million to accelerate environmental approvals for renewable energy, energy transmission and critical minerals projects and deliver additional regional energy transition plans.

But below these headline positive developments a less rosy picture emerges. The Albanese Labor Government's considerable investments in domestic energy transition and manufacturing initiatives have so far not been accompanied by any action to curtail Australia's massive coal and gas exports (MacNeil and Edwards, 2023). The Government's May 2024 *Future Gas Strategy* posits an ongoing role for Australian gas both for domestic use and export beyond 2050 (DISR, 2024) even as the CSIRO's (2024), *GenCost* study, CSIRO, 2024 suggests little need for expansion of gas generation.

Below the national level, state governments have struggled to maintain the momentum of transition amidst growing concern about energy security and affordability. Energy security concerns were behind the NSW Government's decision to underwrite a two-year extension of operation for the Eraring coal-fired power station until August 2027 at a cost of up to $225 million per annum.[3] Budgetary pressures sealed the fate of the Latrobe Valley Authority, which the Victorian Government announced in its May 2024 budget will be absorbed by Regional Development Victoria. At the same time, concerns about the social and environmental impact of new solar and wind generators, electricity transmission lines and large-scale batteries have begun to emerge, particularly in regional communities.

Most recently, the 'climate wars' were fanned back into flame in June 2024 when federal opposition leader Peter Dutton announced that the Coalition would no longer support the Government's emission reduction targets (Morton, 2024). Building on growing community unease about renewable energy expansion plans, Dutton pivoted to announce that a Coalition Government would instead build seven nuclear power stations if elected in 2025. Five of these nuclear power stations would be built in the case study regions discussed in this book: Central Queensland, the Hunter Valley in NSW, the Latrobe Valley in Victoria, Port Augusta in South Australia and Collie in Western Australia (Reimakis and Karp, 2024).

With the possible exception of this latest return to debate about the merits of nuclear energy, recent Australian shifts are in step with international developments. A growing number of governments have flagged reduced support for ambitious emission reductions and transition initiatives amidst growing public concern about energy security and cost of living. As a result, the lessons learned from the Australian regions in this book will find application both domestically and internationally. In the next section, we draw out the key lessons across the five case studies, illustrating them with quotes from each of the chapters. For ease, the quotes from chapters in this book are formatted in *italics*.

Key lessons from Australian regional energy transitions

1. Understanding and respecting the historical and current characteristics of regional communities is vital

The five regional transition journeys explored in this book highlight the fact that all energy transitions are, in the end, local. As such they are shaped by the unique histories, cultures, challenges, concerns, knowledges and experiences of the local communities.

In **Port Augusta, South Australia**, the local community became increasingly focused on the risks arising from the closure of coal-based industries and the opportunities for the region of low-cost renewable energy development. But initial optimism about the potential for community-led change has given way to disappointment in the failure of the State and Commonwealth governments to provide sustained, proactive planning, coordination and resourcing to support the energy transition.

> *Communities need tangible, shared and inspiring visions to become activated for change. When the vision speaks to the unique circumstances of a place, is compatible with local skills and identity, and fosters community pride, it has a better chance of securing support.*
>
> (Chapter 2)

In **Victoria's Latrobe Valley,** workers and communities have been contending with the implications of privatisation since the 1990s. They have learned to be cautious about the regional transition promises and plans of successive State and Commonwealth governments. While the sudden 2017 closure of the Hazelwood coal-fired power station created massive local challenges it also triggered a rapid, concerted response from communities, trade unions and local and state governments to identify and support the growth of new industries, services and employment opportunities.

> *While some elements of the Valley's historical context—including a strong tradition of worker voice and union leadership—helped to enable the transition response, other elements acted to inhibit change. 'Jobs vs environment' narratives associated with local logging and coal industries remain prevalent, while the legacy of privatisation and enduring climate policy uncertainty often still make it difficult to even discuss transition.*
>
> (Chapter 3)

The vast black coal resources of the **Hunter Valley (NSW)** continue to provide employment for thousands of workers in both export-oriented coal mining and coal-fired power generation. Here, it has taken a coordinated, strategically focused coalition of community groups, unions, business and environmental organisations to overcome decades of inertia and begin a conversation about what a shift away from coal means. But significant challenges remain in identifying alternative industries to the still-pervasive coal industry.

> *Coal has permeated the life of the [Hunter] region—forging its working-class character, driving its economy, influencing its*

political economy, scarring its landscape and waters and supporting the livelihoods of many tens of thousands. ... The move from an industrial company town run by Australia's largest company to a more diversified economy disrupted the coherence and influence of regional business, labour and political power structures. The effects of this fragmentation persist to the present day.

(Chapter 5)

The energy transition task in **Gladstone, Central Queensland** has been particularly challenging because fossil fuel production and transportation (especially gas and coal) continue to be central to the regional economy. The success of recent respectful and constructive energy transition dialogues between local communities, workers, businesses and governments illustrates the importance of creating substantial and locally driven engagements.

Change is rarely a neat, linear process. It is more often the outcome of a complex interplay of multiple events—both planned and unexpected. National climate campaigns, ambitious emissions targets among Australia's trading partners and unexpected external shocks such as the COVID pandemic and the 2019 Black Summer Bushfires all contributed to a growing awareness and acceptance that the world was changing, and that Gladstone would have to change with it. What stands out as unique to the Gladstone experience is the pivotal role played by leaders in local government and industry in bringing the community together to host and shape the conversation about transition.

(Chapter 6)

All five case studies highlight the importance of addressing the ongoing legacies of colonial exploitation of First Nations lands and resources. But **Collie, Western Australia** provides a particularly powerful example of how poorly this has been done so far. It highlights that inclusive participation of all segments of community led by First Nations Elders plays a key role in ensuring transitions are just.

Collie's current transition launches from a long history of contestation and activism for competing priorities between Traditional Owners, workers, industry, and environmentalists. ... To date,

> *Collie's transition has missed the opportunity to include Wilman Traditional Owners in transition planning and implementation and incorporate their knowledges, wisdom and cultural responsibilities for Boodja (Country), Bilya (River), Kep (Water), and Moort (Family and People).*
>
> (Chapter 4)

2. Respectful and inclusive engagement with local communities and workers is key

Given the first key lesson, it is not surprising that the second key lesson from our case studies is that listening carefully and respecting local voices, local experience and local leadership is key. The case studies reveal that respectful and inclusive engagement is crucial both because it harnesses local knowledge and experience and because it helps overcome polarisation, creating and sustaining social license and public support for change. This means creating the avenues for local people to co-design decision-making processes as well as participate in them, so they can shape their own futures.

> *... local leaders needed to step forward to champion the need to prepare for transition. These leaders needed support, which they found through engagement approaches which:*
>
> 1. *were designed to meet people where they were at to support them to manage change, rather than push for change;*
> 2. *the leaders themselves were involved in designing; and*
> 3. *highlighted how local processes were connected to bigger national and international transition conversations.*
>
> (Chapter 6)

Consultation, engagement and participation strategies need to be genuinely and *visibly* respectful and inclusive, creating opportunities for a diverse range of voices and points of view. If not, they will be perceived as tokenistic box ticking exercises, which only have the effect of reinforcing cynical perceptions about the motives and assumptions of outsiders.

> *Given long-held local cynicism, and perceptions of power being imposed on the region by 'outsiders' (be they governments,*

> *businesses, activists or others), genuine local participation in transition policies, programs and responses is essential.*
>
> (Chapter 5)

Particular attention needs to be paid to ensuring inclusion of groups who are often sidelined from transition related discussions and planning. Such groups include employees of service industries, subcontractors, low-income households, women, children, and older people as well as those from diverse communities.

> *It is vital that people in all their lived experiences are consistently considered, included, and celebrated in a transition process. Transition planners must engage and communicate with local and diverse voices and leaders and foster equitable and inclusive participation in transition planning and decision-making, as well as provide timely and accessible information, communication and resources about the transition process.*
>
> (Chapter 4)

In many coal communities, engagement with First Nations groups about transition has been limited. Meaningful engagement is not only confounded by the ongoing historical injustices associated with dispossession through colonisation and ongoing industrial development, but also in some cases, by financial ties to the fossil fuel industry through royalty payments.

> *Structural marginalisation of First Nations peoples from transition planning further entrenches the deep injustices and inequalities that First Nations peoples already experience, related to areas such as employment, education, health, housing and incarceration. ... Weak and unfair agreements with native title corporations create lasting community division and can lead to the breakdown of social licence for developers and operators.*
>
> (Chapter 4)

New initiatives such as the First Nations Clean Energy Network, Original Power and the Indigenous Chamber of Commerce are disrupting these patterns by supporting First Nations communities to shape the development of new renewable industries. Supporting First Nations groups to understand how the energy system is changing and how they can get involved is already generating direct economic benefits through community ownership models, jobs, procurement opportunities and even equity arrangements.

Sometimes, the key to respectful and inclusive engagement is simply to remember that the language of 'transition' or 'climate change' might not be the best approach. In some contexts, using the language of 'just transition' has the effect of hindering rather than enabling a productive conversation (Wang and Lo, 2021), while the principles of protecting jobs and livelihoods at all costs can also be used as an argument for preventing and delaying action to phase out fossil fuels (Harry et al., 2024). These lessons were clearly revealed in our case studies, where the terms 'transition' and 'just transition' were frequently seen as toxic and alternatives were suggested such as 'economic diversification' (see also Edwards et al., 2022).

> *Engagement in transition—visioning, planning, critique, execution— doesn't always have to be in the language of climate, energy or even 'transition'. Given the baggage often carried by the term transition, there are times and places where 'talking about transition' can inhibit engagement, action or change.*
>
> (Chapter 3)

3. Proactive, well-coordinated and sustained leadership is needed at all levels

While respectful and inclusive engagement with regional communities is an essential precondition for successful implementation of transition plans, it was clear from our case studies that high-level leadership and co-operation between all levels of government and relevant business, union and community organisations is also vital.

> *The resounding message from Hazelwood's closure is that long-term planning and coordination of transition is essential. With further power station closures scheduled, the region's future depends on ensuring that there is ongoing, well-resourced planning and coordination.*
>
> (Chapter 3)

> *The questions of who speaks for a region and who can to participate are critical in ensuring those who are exposed to change*

> *are supported. Governments therefore must play a central role in ensuring marginalised groups have access to participation.*
>
> (Chapter 5)

Government leadership includes clearly explaining why it is necessary to phase out fossil fuels and developing plausible and compelling implementation plans for the transition process and the new energy economy. It also means adequately resourcing both.

> *The Hunter is now taking tentative but increasingly purposeful steps towards an energy transition. However, translating principle to action through the fog of fear, sectoral lock-in, identity and politics has been challenging. The need for proactive leadership has become apparent as competing interpretations of values such as fairness, and the hard realities of experienced and projected economic change, influence the scope for action.*
>
> (Chapter 5)

The costs and responsibilities of transition strategies need to be transparently and equitably shared. Options for sourcing the funds required to adequately support long-term, comprehensive just transition strategies include general taxation revenue, revenue from a carbon price, revenue recovered by removing fossil fuel subsidies and/or expanding fossil fuel royalties, reverse auction strategies and direct funding from fossil fuel companies to support retraining, infrastructure renewal and the regeneration of mining and power plant sites (Jotzo, 2024).

A common theme across all the initiatives explored in this book is the key role appropriately resourced local, regional and national level transition governance organisations and institutions play in supporting the design and implementation of regional transition strategies.

> *A key lesson from the Hunter's experience to date is that leadership is required to navigate a path through the transition, and that this leadership must be accompanied by institutions that are fit-for-purpose in tackling the long term, structural nature of change.*
>
> (Chapter 5)

> *Many stakeholders in Gladstone have advocated for the establishment of a regional transition authority or committee to oversee the*

> *implementation and coordination of transition related initiatives. ... While local leaders are doing what they can, without an institution that has the authority to bring all stakeholders together to regularly review plans and progress, many are now reporting that there is an increasing lack of coordination as the pace of change accelerates.*
>
> (Chapter 6)

Local grassroots community organisations like Voices of the Valley, Repower Port Augusta and the Hunter Jobs Alliance have played vital roles in creating safe spaces in which communities in our case study areas could share concerns and explore alternative economic pathways.

> *Local leadership—working in formal and informal spaces—has been a pivotal feature in the [Latrobe] Valley's transition to date. From the early leadership of the GLTC [Gippsland Trades and Labor Council] on regional diversification and just transition strategies to the community organising and education of VOTV [Voices of the Valley] and GCCN [Gippsland Climate Change Network] and the economic experimentation of Earthworker, there is a long tradition of local groups and individuals facilitating challenging conversations and leading change.*
>
> (Chapter 3)

Several of the case studies also highlight the challenges local governments face in addressing the impact of coal mine and power plant closure decisions as well as in co-designing and supporting complex economic diversification and renewal plans and strategies (Rural Network, 2024). This is particularly the case where they have limited powers and resources at their disposal.

> *Council capacity to lead, manage and derive value from renewables is central to success. Councils are at the front line in regional communities. When renewable energy companies aren't required to pay appropriate Council rates (or equivalent) the social licence of the industry is undermined. State and federal governments need to entrust councils with more autonomy and resource their capacity to respond to concerns and manage energy transition well.*
>
> (Chapter 2)

The Port Augusta case study highlighted the fact that local government also plays a key role in ensuring governments and companies stay focused on the impacts on communities throughout all stages of the transition process, including the development of new industries as well as the phase out of the old:

> *Councils also need negotiating power that encourages renewable companies to foster and maintain relationships with them throughout the project development, operational and remediation stages.*
>
> (Chapter 2)

4. Well-planned, adequately funded re-employment, retraining and early retirement programmes are needed for workers

Workers in the coal-fired power sector at risk of losing their jobs need to be adequately supported through a coordinated programme of re-employment, retraining, relocation, income support and early retirement programmes appropriate to different ages, aspirations and career stages. As noted in Chapters 2 and 5:

> *long-term skilled work in the renewable construction industry mostly involves years of specialist training, meaning large contractors tend to use FIFO models for construction work. The work available to locals living near a construction project is therefore mostly lower skilled and short term.*
>
> (Chapter 2)

> *Practical actions such as subsidised training for new and sustainable industries, worker transfer schemes, and high-quality labour market services for affected workers have all been highlighted by labour unions and community organisations as being crucial to a fair energy transition in the Hunter.*
>
> (Chapter 5)

In this regard, recent international experience demonstrates the value of proactive, carefully coordinated pooled redundancy schemes in enabling workers from plants scheduled for closure to be offered redeployment opportunities at other power stations which are continuing to operate.

> *Diverse groups and long-term unemployed locals face particular barriers to employment within a high pressure and profit driven employment environment. Systematic change, advanced planning, contractor coordination and practical wrap-around worker support is required to enable these groups to access employment in the renewable construction industry.*
>
> (Chapter 2)

But new clean energy jobs will not necessarily be in the same locations as existing fossil fuel sector jobs. Some workers will need to seek employment in new locations, so relocation and retraining packages are also needed. Opportunities to access retraining should be made available prior to closures and redundancies so workers can access them while still earning an income.

> *To achieve new, clean industries with sustainable jobs for existing and future workers, transitions must foster extensive education and training opportunities that are high quality, accessible, and affordable for community members of all ages and diversities. This might include retraining, one-to-one support, career guidance, and counselling for workers and businesses to diversify.*
>
> (Chapter 4)

5. Transition planning must promote economic renewal and create high-quality jobs

As the UNFCCC *Just Transitions Report* (2022) notes, a clean energy economy will only provide positive employment impacts overall if the transition addresses the concerns of workers and communities who are unable to easily access new work opportunities. 'Regions which lack diversification, which have a limited capacity for innovation, or whose economic mainstay is vulnerable to decisions made elsewhere will face the greatest challenge, as will workers with skills that are in less demand or who are unable to acquire new skills' (UNFCCC, 2022).

Our case studies show that this requires proactive efforts.

> *Whether focused on assuring the future of emissions-exposed industries with decarbonisation potential, seeking a share of global clean energy industry opportunities, or adjusting to structurally changed economies by creating 'lots of little industries*

everywhere,' there is recognition in the Hunter that more proactive efforts are required.

(Chapter 5)

In large part, this reflects the fact that creating secure high-quality replacement jobs can be very challenging, but that public and political support for transition depends on communities being convinced that governments and business are genuinely committed to trying.

Establishing the expectations and means to treat workers fairly through change is half the battle. The other half of the economic transition equation—finding ways to replace economic activity and jobs—is incredibly difficult. ... There is a fundamental challenge in reconciling community expectations for 'traditional' blue collar work with the realities of a modern service-based economy where such work is proportionally diminishing.

(Chapter 5)

Research on the economic renewal challenges facing coal dependent regions highlights the tendency for their regional economies to be overspecialised in declining, capital-intensive industries. Ownership of incumbent industries by a small number of large (often multi-national) companies often creates high barriers to entry for new entrepreneurs and start-ups. Our case studies showed that communities often fall into the trap of looking for 'silver bullet' solutions in which a single, large-scale industry successfully replaces the economic and employment outcomes previously provided by coal. For instance:

Many in the [Latrobe] Valley seek another major industry to arrive and 'save' the region by essentially replacing one industrial employer for another near identical one. This regional development strategy of 'chasing smokestacks' to 'attract the golden egg of corporate employment' is rarely successful.

(Chapter 3)

There is also increasing awareness that long-term economic value for communities hosting renewables is more likely to be found by developing local and international supply chain opportunities and by encouraging new manufacturing, energy and labour-intensive industries to co-locate with renewables than by attempting to directly substitute the old energy industries with the new:

> *Those who approach the Valley's challenge as primarily an energy transition are therefore commonly stuck seeking an impossibly neat swap from local coal energy jobs to equal quantities of local renewable energy jobs. Electricity generation is certainly part of the Valley's transition, but transition in the Valley is necessarily a question of diversification, both of economic activities and regional identities.*
>
> (Chapter 3)

'Smart specialisation' strategies and policies have the potential to assist in overcoming industrial path dependency and encouraging economic diversification. According to recent research, these strategies can include:

- the identification and amplification of regional strengths and assets, informed by local knowledge and expertise;
- prioritising a culture of locally driven innovation and entrepreneurialism over short-term protection for declining industries;
- identifying and enabling vertical and horizontal industry and supply chain linkages and collaboration;
- facilitating new job creating innovation and investment including through diversification and new market access advice, new product and process commercialisation support, tax and co-investment incentives, business loan guarantees; and regional procurement strategies; and
- encouraging university, TAFE and other research institute partnerships to support new start-up businesses (Miedzinski et al., 2022).

To implement such strategies requires decision-makers to extend the scope of energy transition conversations well beyond debates about likely local renewable energy opportunities to the geographical scale of the transition project.

> *Defining the scale of transition, particularly a 'just transition', is a persistent problem in theory and practice. Is the transition solely*

> *to replace the Valley's dwindling brown coal industry, or should it include the wider Gippsland region as native timber and oil and gas industries also decline, and entrenched socio-economic disadvantage persists?*
>
> (Chapter 3)

In this sense, our case studies confirm that integrated, government led industrial policy plays a significant role in framing such conversations. Examples are beginning to emerge, including the US Inflation Reduction Act, the European Green New Deal and the recently announced Future Made in Australia initiative. Each of these create valuable foundations for the design and implementation of regional transition plans and strategies.

> *One critical shift has been a renewed government emphasis on regional industry policy that has made the Hunter a national target for job-generating public investments in the clean energy economy. While the effectiveness of such investments at both national and regional levels are yet to be seen ... the substantial public investment infrastructure being assembled is encouraging.*
>
> (Chapter 5)

But coordination alone is insufficient. Jobs created in the transition to a low-carbon economy must also be 'decent'. This means jobs that provide adequate incomes, safe working conditions, respect for rights at work and social protections. As the ILO reminds and cautions:

> managed well, transitions to environmentally and socially sustainable economies can become a strong driver of job creation, job upgrading, social justice and poverty eradication… [but] the job-creating potential of environmental sustainability is not a given: the right policies are needed to promote green industries while ensuring decent work within them.
>
> (ILO, 2015)

To achieve employment outcomes on this scale and of this quality requires strong political leadership and sustained strategic investment in infrastructure, supply chain logistics, research, education and training and labour market programmes.

6. Transition always affects the whole community and attention to the most marginalised is essential

Our case studies consistently highlight that attention needs to be paid to addressing the employment and community support needs of vulnerable individuals including women, low-income households, First Nations communities and other marginalised groups.

> *Transition planning which takes a narrow focus on usually male-dominated industries such as manufacturing, construction and emergency services will tend to neglect broader social justice issues that the community wants to see addressed. In Collie's case, this includes the degradation of Country, access to quality housing, social services and physical and mental health support, and access to appropriate childcare and aged care.*
>
> (Chapter 4)

Communities, as noted in Chapter 2, can also 'play a powerful role in attracting the industries they want to their region, including through committed volunteers, financial resourcing, proactive leadership, alliances of organisations and politicians who champion their goals'.

A key lesson from our case studies, then, is that the impact of energy transitions always extends well beyond the workers directly employed in plants and mines earmarked for closure. Contractors, supporting businesses, households and communities are all affected and all need support, particularly women who bear the brunt of substance abuse and violence (Wiseman and Wollersheim, 2021). This was highlighted in Collie, which found that:

> *Support for vulnerable populations is key to a successful transition. This ranges from infrastructure such as affordable public transport, accessible recreation facilities, and social and affordable housing, to well-funded services in areas such as aged care, physical andmental healthcare, disability support, and childcare. There is a particular deep need for mental health support for affected workers, families and the broader community through the transition process and impacts of climate change, to strengthen community resilience and sustainability.*
>
> (Chapter 4)

> *Fundamental to this is ensuring that the transition supports all community members to participate in work, by addressing structural barriers such as caring responsibilities and racism. This includes ensuring equitable access to childcare and addressing barriers to employment for workers with disability or workers caring for people with disability or other needs.*
>
> (Chapter 4)

As Samantha Smith, co-ordinator of the International Trade Union Confederation's Just Transition Centre notes, reversing the vicious cycle of deindustrialisation and revitalising local community economies requires careful planning and sustained effort: 'It also takes investment in infrastructure, public services, schools and training facilities and hospitals—in short, all of the things that draw employers and families back to the region' (Smith, 2017).

Broadening the focus of transition strategies in this way also highlights the potential for these strategies to assist communities to meet the challenges of increasingly frequent, increasingly severe climate-related fires, floods, storms and heatwaves.

> *economic transition is an opportunity to respond to the changing climate and reduce emissions along with supporting just adaptation and enhancing community resilience to climate change and disasters. New policies, public expenditure, industries and technologies can help facilitate climate justice, while climate resilience also requires coordinated disaster prevention, preparedness, response and recovery from a social justice lens.*
>
> (Chapter 4)

7. The environmental aspects of the transition must not be neglected

Finally, though it was not generally at the forefront of our case studies, the environmental remediation of industrial land and mine sites must be adequately resourced. Governments need to ensure that coal mining and power generation companies allocate sufficient funds to support site rehabilitation and transition strategies before they relocate or go bankrupt.

> *A transition from coal mining across large parts of the landscape presents an opportunity for a revised approach. ... It is damaged*

> *but can be restored. Conversely, failing to deliver ecological justice will undermine the broader prospects of a just transition by unnecessarily alienating a significant section of the community who are concerned about the mining industry's landscape impacts and encouraged by the prospects of long-term landscape restoration.*
>
> (Chapter 5)

In this sense, the transition is always an opportunity to recast the relationship between community and nature in positive ways.

> *While the ecological pieces can never be put together the same way again, there is an opportunity to create a restorative, and potentially world leading, environmental outcome. Addressing environmental policy failures requires the same leadership and institutional innovation demanded by the economic transition.*
>
> (Chapter 5)

Strengthening and accelerating just transitions: key challenges, opportunities and priorities

In this chapter, we have highlighted seven key lessons from our five case study regions, each of which had already begun to confront the challenges and opportunities of a just and rapid energy transition before the political climate in Australia changed in 2022. The case studies show that every transition is geographically and historically specific. This means that the lessons learned in one place cannot simply be transferred to another or scaled up into a general principle. But some themes have emerged about what drives success and what challenges must be overcome.

For regional energy transitions to be just, they must provide for the workers who will be most directly affected. This should take the form of well-planned, adequately funded re-employment, retraining and early retirement programmes for those working in the fossil fuel industries that are reaching their sunset. Transitions must also promote new high-quality jobs and economic renewal which benefits all members of the community, including those who are currently marginalised and whose histories have been punctuated by compounding disadvantage. But initiatives that address only the employment dimension of the transition will fail to be truly just, regardless of the degree to which they seek to create employment for marginalised groups. Our case studies have consistently pointed to the fact that just transition only becomes possible when local communities are respectfully and inclusively engaged in the conversation. Engagement does not merely mean informing or consulting communities; it means a much more robust process of listening, co-design and direct involvement throughout the transition.

But arguably, the single most significant lesson from these regions collectively is simply that it is possible to put transition on the agenda even when politically and practically it seems impossible. What it takes to do that is leadership—particularly local leadership—which is focused on building consensus amongst actors with different primary motivations. These actors range from industry leaders to environmental and social groups. An essential precondition for this leadership is the willingness to have respectful but frank conversations about what are often uncomfortable realities.

Australia has begun to establish transition authorities and bodies at different levels, and it is becoming clear that they are a crucial part of the puzzle—particularly to coordinate funding and implementation of transition policies across different regions—but will be far from sufficient to promote a truly just transition alone. A key challenge is to ensure that they engage effectively with local communities to build on regional aspirations, knowledge and strengths. Transition requires the availability of skilled workers to build, maintain and run the infrastructure of the new low-carbon economy, but *just transition* requires including and lifting up marginalised and vulnerable communities. So far in Australia, the signs are not promising that the transition will overcome existing structural disadvantage. The early signs are that as the transition accelerates it will fail to adequately engage with and learn from both women and First Nations people, to its great detriment.

While just regional energy transitions have moved into the realm of possibility in Australia, they are certainly not guaranteed. As the energy transition progresses, it is critical that the phaseout of fossil fuels continues to accelerate. Justice will be critical to ensuring the durability of transitions in fossil fuel-producing regions and overcoming the economic and political challenges that drive ongoing fossil fuel production. This entails addressing both the supply of fossil fuels and demand for them, often in different places. There will be no just transition overall if the principles of 'just transition' are only applied in isolation in particular communities. Attention must also be paid to the connection between these communities and the broader national and international contexts they are located in.

As noted in Chapter 1, increasing recognition of the speed with which fossil fuels need to be phased out combined with sharp falls in renewable energy costs are creating both opportunities and challenges for regional communities with a long history of reliance on fossil fuel industries. This is the case both in Australia and overseas. In this context, the recent Australian regional energy transition initiatives examined in this book teach valuable lessons to both policymakers and researchers. Our hope is that this book catalyses an ongoing conversation about how to maximise the social, environmental and economic opportunities of the energy transition which is already underway. It is critically important that this transition is just, inclusive and well managed, and that the new society and economy it is ushering in is just and regenerative as well as zero-carbon.

Notes

1 www.aph.gov.au/Parliamentary_Business/Bills_Legislation/Bills_Search_Results/Result?bId=r7177?>
2 https://budget.gov.au/content/03-future-made.htm?>
3 www.nsw.gov.au/media-releases/nsw-government-secures-two-year-extension-to-eraring-power-station

References

Ciplet D and Harrison JL (2020) Transition tensions: Mapping conflicts in movements for a just and sustainable transition. *Environmental Politics* 29(3): 435–456.

CSIRO (2024) GenCost 2023–2024. www.csiro.au/en/research/technology-space/energy/gencost (accessed 22 July 2024).

Department of Industry, Science and Resources (2024) Future gas strategy. www.industry.gov.au/publications/future-gas-strategy (accessed 22 July 2024).

Edwards GAS, Hanmer C, Park S, MacNeil R, Bojovic M, Kucic-Riker J, Musil D and Viney G (2022) *Towards a just transition from coal in Australia?* London: The British Academy. https://doi.org/10.5871/just-transitions-a-p/G-E

Harry SJ, Maltby T and Szulecki K (2024) Contesting just transitions: Climate delay and the contradictions of labour environmentalism. *Political Geography* 112: 103114.

ILO (2015) *Guidelines for a just transition towards environmentally sustainable economies and societies for all.* Geneva: International Labour Organisation.

Jotzo F (2024) How is the transition to net zero going to be paid for? *Australian Financial Review*, June 3. www.afr.com/policy/energy-and-climate/how-is-the-transition-to-net-zero-going-to-be-paid-for-20240602-p5jij6 (accessed 22 July 2024).

MacNeil R and Edwards GAS (2023) The promise and peril of Australian climate leadership under Albanese. *Australian Journal of International Affairs* 77(1): 19–25.

Miedzinski M, Coenen L, Larsen H, Matusiak M and Sarcina A (2022) Enhancing the sustainability dimension in smart specialisation strategies: A framework for reflection. European Commission. https://lva.vic.gov.au/transition/aligning-smart-specialisation-with-sustainable-development-goals/enhancing-the-sustainability-dimension-in-smart-specialisation-KJNA31322ENN.pdf (accessed 22 July 2024).

Morton A (2024) Peter Dutton accused of trying to 'rip up' Australia's commitment to Paris Climate Agreement. *The Guardian*, 8 June. www.theguardian.com/environment/article/2024/jun/08/peter-dutton-accused-of-trying-to-rip-up-australias-commitment-to-paris-climate-agreement (accessed 24 June 2024).

Remeikis A. and Karp P (2024) Peter Dutton names several potential nuclear power station sites. *The Guardian*, 19 June. www.theguardian.com/environment/article/2024/jun/19/coalition-nuclear-plan-peter-dutton-power-station-sites-australia (accessed 24 June 2024).

Rural Network (2024) No one understands local issues better: Rural councils call for greater role in energy transition. *The Guardian*, 22 July. www.theguardian.com/australia-news/article/2024/jul/22/renewable-energy-transition-rural-australia-councils (accessed 22 July 2024).

Smith S (2017) *Just transition: A report for the OECD.* ITUC, May 2017.

UNFCCC (2022) Just transition of the workforce. https://unfccc.int/sites/default/files/resource/Just%20transition.pdf (accessed 22 July 2024).

Wang X and Lo K (2021) Just transition: A conceptual review. *Energy Research & Social Science* 82: 102291.

Weller S, Beer A and Porter J (2024) Place-based just transition: Domains, components and costs. *Contemporary Social Science 19*(1–3): 355–374. https://doi.org/10.1080/21582041.2024.2333272

Wiseman J and Wollersheim L (2021) Building just and resilient zero carbon regions. *Melbourne Climate Futures*. www.unimelb.edu.au/__data/assets/pdf_file/0009/3934404/Wiseman-and-Wollersheim,-2021_MCF-Discussion-Paper_final.pdf (accessed June 17 2024).

Index

Note: Figures are shown in *italics*. Endnotes are indicated by the page number followed by "n" and the endnote number e.g., 139n1 refers to endnote number 1 on page 139.